Programming the Raspberry Pi™

About the Author

Dr. Simon Monk has a bachelor's degree in cybernetics and computer science and a Ph.D. in software engineering. He is now a full-time writer and has authored numerous books, including *Programming Arduino, 30 Arduino Projects for the Evil Genius, Hacking Electronics,* and *Raspberry Pi Cookbook.* Dr. Monk also designs products for MonkMakes.com. You can follow him on Twitter, where he is @simonmonk2.

Programming the Raspberry Pi™

Getting Started with Python

THIRD EDITION

Simon Monk

New York Chicago San Francisco
Athens London Madrid
Mexico City Milan New Delhi
Singapore Sydney Toronto

Library of Congress Control Number: 2021935766

McGraw Hill books are available at special quantity discounts to use as premiums and sales promotions or for use in corporate training programs. To contact a representative, please visit the Contact Us page at www.mhprofessional.com.

Programming the Raspberry Pi™: Getting Started with Python, Third Edition

1 2 3 4 5 6 7 8 9 LCR 26 25 24 23 22 21

ISBN 978-1-264-25735-5
MHID 1-264-25735-X

This book is printed on acid-free paper.

Sponsoring Editor
Lara Zoble

Editorial Supervisor
Stephen M. Smith

Production Supervisor
Lynn M. Messina

Acquisitions Coordinator
Elizabeth M. Houde

Project Manager
Jyoti Shaw, MPS Limited

Copy Editor
Md. Taiyab Khan, MPS Limited

Proofreader
Kirti Dogra, MPS Limited

Indexer
Mary Kidd

Art Director, Cover
Jeff Weeks

Illustration
MPS Limited

Composition
MPS Limited

To my brothers, Andrew and Tim Monk, for their love and wisdom.

CONTENTS AT A GLANCE

CONTENTS

PREFACE

The Raspberry Pi™ is rapidly becoming a worldwide phenomenon. People are waking up to the possibility of a $35 (U.S.) computer that can be put to use in all sorts of settings—from a desktop workstation to a media center to a controller for a home automation system.

This book explains in simple terms, to both nonprogrammers and programmers new to the Raspberry Pi, how to start writing programs for the Pi in the popular Python programming language. It then goes on to give you the basics of creating graphical user interfaces and simple games using the `pygame` module.

The software in the book uses Python 3, and the Mu editor. The Raspberry Pi OS distribution recommended by the Raspberry Pi Foundation is used throughout the book.

The book starts with an introduction to the Raspberry Pi and covers the topics of buying the necessary accessories and setting everything up. You then get an introduction to programming while you gradually work your way through the next few chapters. Concepts are illustrated with sample applications that will get you started programming your Raspberry Pi.

Four chapters are devoted to programming and using the Raspberry Pi's GPIO connector, which allows the device to be attached to external electronics. These chapters include three sample projects—a LED lighting controller, a LED clock, and a Raspberry Pi–controlled robot, complete with ultrasonic rangefinder.

Here are the key topics covered in the book:

- Python numbers, variables, and other basic concepts
- Strings, lists, dictionaries, and other Python data structures
- Modules and object orientation
- Files and the Internet
- Graphical user interfaces using guizero

- Game programming using pygame
- Interfacing with hardware via the GPIO connector
- Sample hardware projects

All the code listings in the book are available for download from the book's repository on Github at https://github.com/simonmonk/prog_pi_ed3, where you can also find other useful material relating to the book, including errata.

Simon Monk

ACKNOWLEDGMENTS

As always, I thank Linda for her patience and support.

At TAB/McGraw Hill, my thanks go out to my editor Lara Zoble, and I also thank Jyoti Shaw of MPS Limited. As always, it was a pleasure working with such a great team.

INTRODUCTION

Since the first Raspberry Pi™ model B revision 1 was released in 2012, there have been a number of upgrades to the original hardware. The Raspberry Pi 4 has increased the processing power and memory of the Raspberry Pi and the Pi Zero provides a very low cost option, while the Raspberry Pi 400 is actually built into a keyboard. These new versions of the Raspberry Pi have been largely compatible with the original device, but there are a few changes to both the hardware and the standard Raspberry Pi OS distribution that warrant a new edition of this book.

In particular, I have changed all the user interface code from Tkinter to the much easier to use guizero and I have also changed the code examples that used RPi.GPIO to gpiozero.

The Raspberry Pi Zero.

Much of this book is concerned with learning Python, the most common programming language used with the Raspberry Pi, and this remains largely unchanged. However, Chapter 7 has been rewritten to use guizero, and Chapters 9 to 11, which deal with hardware, have been updated to use the gpiozero library.

Although, at the time of writing this book, the current model of Raspberry Pi is the Raspberry Pi 4, for simplicity I will just use the term Raspberry Pi to refer to all models of Raspberry Pi unless the situation needs a distinction to be drawn.

1

Introduction

The Raspberry Pi™ went on general sale at the end of February 2012 and immediately crashed the websites of the suppliers chosen to take orders for it.

Since then a number of new models culminating in the Raspberry Pi 4 (at the time of writing) have been released. So what was so special about this little device and why has it created so much interest?

What Is the Raspberry Pi?

The Raspberry Pi 4, shown in Figure 1-1, is a computer that runs the Linux operating system. It has USB sockets you can plug a keyboard and mouse into and no less than two HDMI (High-Definition Multimedia Interface) video outputs you can connect a TV or monitor into. Many monitors only have a VGA connector, and Raspberry Pi will not work with this. However, if your monitor has a DVI connector, cheap HDMI-to-DVI adapters are available.

When Raspberry Pi boots up, you get the Linux desktop shown in Figure 1-2. This really is a proper computer, able to run an office suite, video playback capabilities, games, and the lot. It's not Microsoft Windows; instead, it is Windows' open source rival Linux (Debian Linux), and the windowing environment is called Pixel.

It's small (the size of a credit card) and extremely affordable (starting at $30). Part of the reason for this low cost is that some components are not included with the board or are optional extras. For instance, it does not come in a case to protect it—it is just a bare board. Nor does it come with a power supply, so you will need to find yourself a 5V USB-C power supply, much like you would use to charge a phone (the power supply capable of supplying 2A and 3A is recommended). Note

Figure 1-1 *The Raspberry Pi 4.*

Figure 1-2 *The Raspberry Pi Pixel desktop.*

that previous models of Raspberry Pi use a micro-USB connector for power rather than USB-C. They also use less current.

What Can You Do with a Raspberry Pi?

You can do pretty much anything on a Raspberry Pi that you can on any other Linux desktop computer, with a few limitations. The Raspberry Pi uses a micro-SD card in place of a hard disk. The older Raspberry Pi models A and B use a full-size SD card, although you can plug in a USB hard disk. You can edit office documents, browse the Internet, and play games (even games with quite intensive graphics, such as *Quake*).

The low price of the Raspberry Pi means that it is also a prime candidate for use as a media center. It can play on two screens at 4k resolution.

A Tour of the Raspberry Pi

Figure 1-3 labels the various parts of a Raspberry Pi. This figure takes you on a tour of the Raspberry Pi 4.

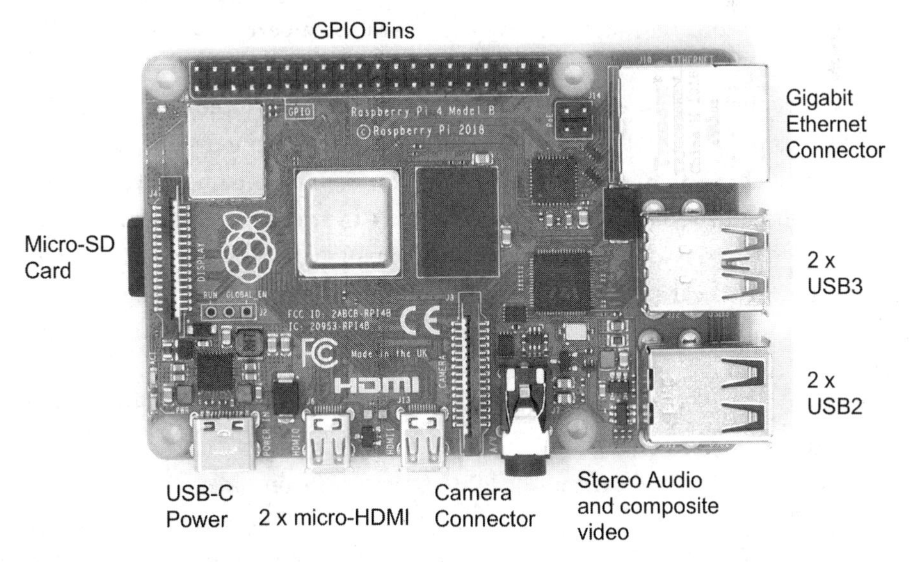

Figure 1-3 *The anatomy of a Raspberry Pi 4.*

The RJ-45 Ethernet connector is shown in the top-right corner of the figure. If your home hub is handy, you can plug your Raspberry Pi directly into your local network. The Raspberry Pi 4 and even earlier models like the Raspberry Pi 3 have built-in Wi-Fi that is usually a more convenient way of connecting to your network and the Internet.

Immediately below the Ethernet socket you'll find two pairs of USB sockets. You can plug a keyboard, mouse, or external hard disks into the board.

In the bottom-center of the figure you'll find an audio socket that provides a stereo analog signal for headphones or powered speakers. This socket also provides a composite video signal. The HDMI connector is also sound capable.

You are unlikely to use the composite video feature of the audio/AV socket connector unless you are using your Raspberry Pi with an older TV. You are far more likely to use one of the HDMI connectors. HDMI is higher quality, includes sound, and can be connected to DVI-equipped monitors with a cheap adapter.

At the top of the Pi are two rows of pins. These are called GPIO (General Purpose Input/Output) pins, and they allow the Raspberry Pi to be connected to custom electronics. Users of the Arduino and other microcontroller boards will be used to the idea of GPIO pins. Later, in Chapter 12, we will use these pins to enable our Raspberry Pi to be the "brain" of a little roving robot by controlling its motors. In Chapter 11, we will use the Raspberry Pi to make an LED clock.

The Raspberry Pi 2 has a micro-SD card slot underneath the board. This SD card needs to be at least 8GB in size. It contains the computer's operating system as well as the file system in which you can store any documents you create, so it's a good idea to get a bigger SD card than the minimum. 32GB is a good size. The SD card is an optional extra when buying your Raspberry Pi. Preparing your own SD card is a little unusual, and suppliers such as SK Pang, Farnell, and RS Components all sell already-prepared micro-SD cards. Because no disk is built into your Raspberry Pi, this card is effectively your computer, so you could take it out and put it in a different Raspberry Pi and all your stuff would be there.

Below the micro-SD card is a USB-C (micro-USB on older Raspberry Pis) socket. This is used to supply power to the Raspberry Pi. Therefore, you will need a power supply with a USB-C on the end. This is the same type of connector used by many mobile phones, including most Android phones. Do, however, check that it is capable of supplying at least 2.5A; otherwise, your Raspberry Pi may behave erratically.

For those interested in technical specs, the big square chip in the center of the board is where all the action occurs. This is Broadcom's "System on a Chip" and includes 1, 4, or 8GB (depending on your Pi 4) of memory as well as the graphics and general-purpose processors that drive the Raspberry Pi 4.

You may also have noticed flat cable connectors on the Pi 4. The connector on the far left is for an LCD display and the connector bottom-center is for the special Raspberry Pi Camera Module.

Setting Up Your Raspberry Pi

You can make your life easier by buying a prepared micro-SD card and power supply when you buy your Raspberry Pi, and for that matter you may as well get a USB keyboard and mouse (unless you have them lurking around the house somewhere). Let's start the setup process by looking at what you will need and where to get it from.

Buying What You Need

Table 1-1 shows what you will need for a fully functioning Raspberry Pi 4 system. The Raspberry Pi itself is sold through two worldwide distributors based in the United Kingdom: Farnell (and the related U.S. company Newark) and RS Components, as well as many online hobby electronics companies like Adafruit and Sparkfun.

Power Supply

Figure 1-4 shows a typical USB power supply.

You may be able to use a power supply from an old phone or the like, as long as it is 5V and can supply enough current. It is important not to overload the power supply because it could get hot and fail (or even be a fire hazard). Therefore, the power supply should be able to supply at least 2.5A, but 3A would give the Raspberry Pi a little extra when it comes to powering the devices attached to its USB ports. If you have an older model B Pi 2 or 3, then a 1.5A micro-USB power adapter will be sufficient.

Item	Source and Part Number	Additional Information
USB power supply (U.S. plug)	PiShop.us: 1660 Buyaoi.ca: 1660 Adafruit: 4298	5V USB power supply. For a Raspberry Pi 4, 3A (15W) is recommended.
USB power supply (UK plug)	Pimoroni.co.uk: RPI040 cpc.farnell.com: SC15228	
Keyboard and mouse	Any computer store	Any USB keyboard will do. Also, wireless keyboards and mice that come with their own USB adaptor will work too.
TV/monitor with HDMI	Any computer/electrical store	
Micro-HDMI to HDMI lead	Any computer/electrical store	
Micro-SD card (32GB class 10 recommended)	Any computer/electrical store	
Ethernet patch cable*	Any computer store	
Case*	Any reseller of Raspberry Pis, also Amazon and eBay	Make sure the case you order is compatible with your model of Raspberry Pi. A Raspberry Pi 4 will not fit in a Raspberry Pi 3 case.
** These items are optional.*		

Table 1-1 *A Raspberry Pi Kit*

Figure 1-4 *USB power supply.*

If you look closely at the specs written on the power supply, you should be able to determine its current supply capabilities. Sometimes its power-handling capabilities will be expressed in watts (W); if that's the case, it should be 15W, this is equivalent to 3A.

Keyboard and Mouse

The Raspberry Pi will work with pretty much any USB keyboard and mouse. You can also use most wireless USB keyboards and mice—the kind that come with their own dongle to plug into the USB port. This is quite a good idea, especially if they come as a pair. That way, you are only using up one of the USB ports. This will also come in quite handy in Chapter 11 when we use a wireless keyboard to control our Raspberry Pi–based robot. If you are using a Pi Zero, you will also need a USB on-the-go to full-size USB adapter.

Display

A low-cost 22-inch LCD TV with a HDMI socket makes a perfectly adequate display for the Pi. Indeed, you may just decide to use the main family TV, just plugging the Pi into the TV when you need it.

If you have a computer monitor with just a VGA connector, you are not going to be able to use it without an expensive converter box. On the other hand, if your monitor has a DVI connector, an inexpensive adapter will do the job well.

Micro-SD Card

You can use your own micro-SD card in the Raspberry Pi, but it will need to be prepared with the NOOBS (New Out of the Box Software) installer. This is a little fiddly, so you may just want to spend a dollar or two more and buy a micro-SD card that is already prepared and ready to go. Most places that sell a Raspberry Pi will also sell ready formatted micro-SD cards with NOOBS pre-installed.

You can also find people at Raspberry Pi meet-ups who will be happy to help you prepare a micro-SD card. Look around on the Internet to find suppliers who sell prepared cards, with NOOBS. If you indeed want to "roll your own" SD card, refer to the instructions found at www.raspberrypi.org/downloads.

To prepare your own card, you must have another computer with a SD card reader.

Case

The Raspberry Pi does not come in any kind of enclosure. This helps to keep the price down, but also makes it rather vulnerable to breakage. Therefore, it is a good idea to buy a case as soon as you can. Figure 1-5 shows a few of the ready-made cases currently available.

Figure 1-5 *Commercial Raspberry Pi cases.*

Figure 1-6 *A homemade Raspberry Pi case.*

You should find a selection of cases available from wherever you bought your Raspberry Pi. Some cases for the Raspberry Pi 4 come with an integral fan. This is not a bad idea, if you plan to work your Raspberry Pi hard, say as a media center.

If you have a 3D printer, take a look at thingiverse.com where you will find no end of Raspberry Pi enclosure designs to print.

People are having a lot of fun building their Raspberry Pi into all sorts of repurposed containers, such as vintage computers and games consoles. One could even build a case using Legos. My first case for a Raspberry Pi was made by cutting holes in a plastic container that used to hold business cards (see Figure 1-6).

Connecting Everything Together

Now that you have all the parts you need, let's get it all plugged together and boot your Raspberry Pi for the first time. Figure 1-7 shows how everything needs to be connected. Setup is easier if your Raspberry Pi is connected to the Internet, which you can either do using the built-in Wi-Fi or an Ethernet cable to your Home Hub.

Insert the micro-SD card with NOOBS, connect the keyboard, mouse, and monitor to the Pi, attach the power supply, and you are ready to go.

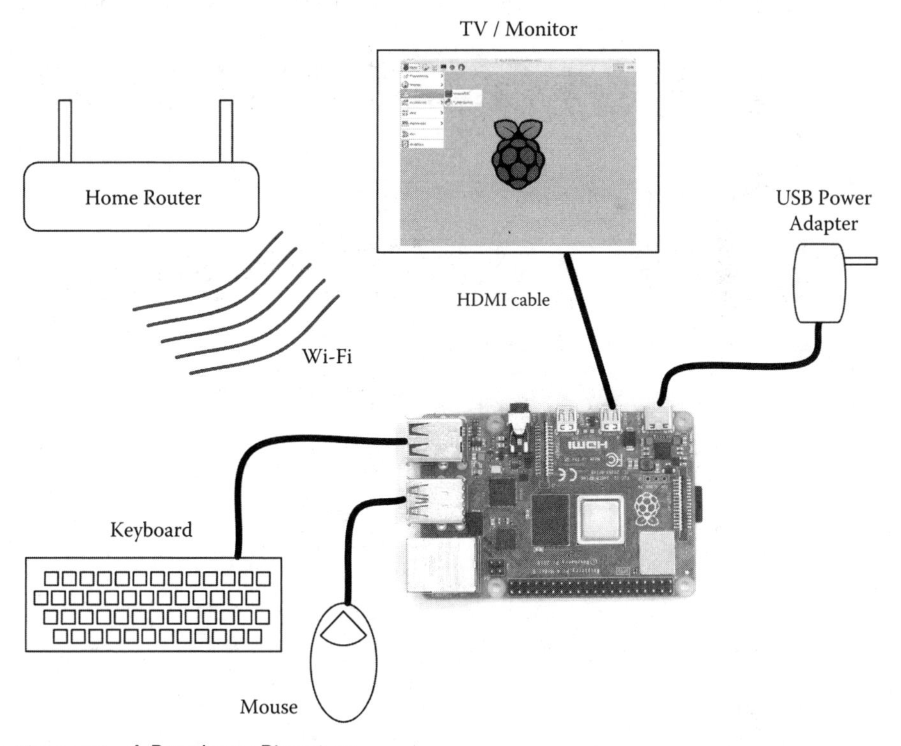

Figure 1-7 *A Raspberry Pi system.*

Booting Up

To make sure that your installer will get the latest version of Raspberry Pi OS, you will connect your Raspberry Pi to your network during the installation process.

When the Raspberry Pi boots into the NOOBS Installer, you will be presented with a list of operating systems (Figure 1-8). Click the checkbox next to the first option (Raspberry Pi OS [RECOMMENDED]) and then click on the "Install" button.

The SD card setup procedure for Raspberry Pi changes from time to time. For latest information on this, please refer to: https://www.raspberrypi.org/documentation/installation/

After a warning that everything on the SD card will be erased, the installation will begin. During this process, which takes quite a while, the installer will show a series of informative messages (Figure 1-9).

When the installer has finished installing Raspberry Pi OS, an alert will pop up to tell you that installation has and ask you to connect to a Wi-Fi network (Figure 1-10).

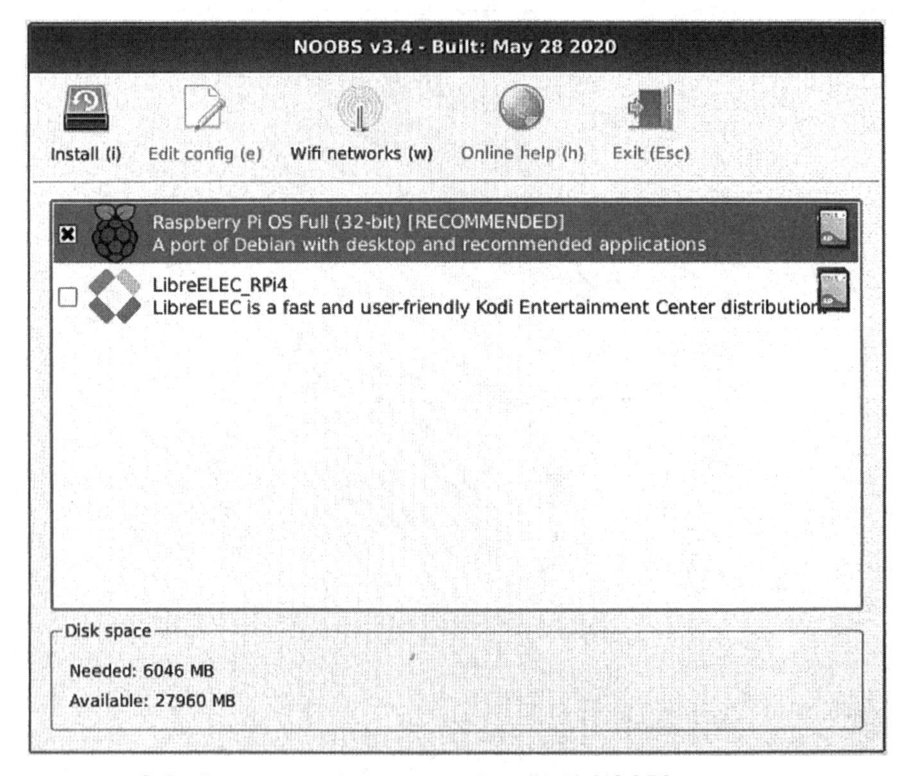

Figure 1-8 *Selecting an operating system to install with NOOBS.*

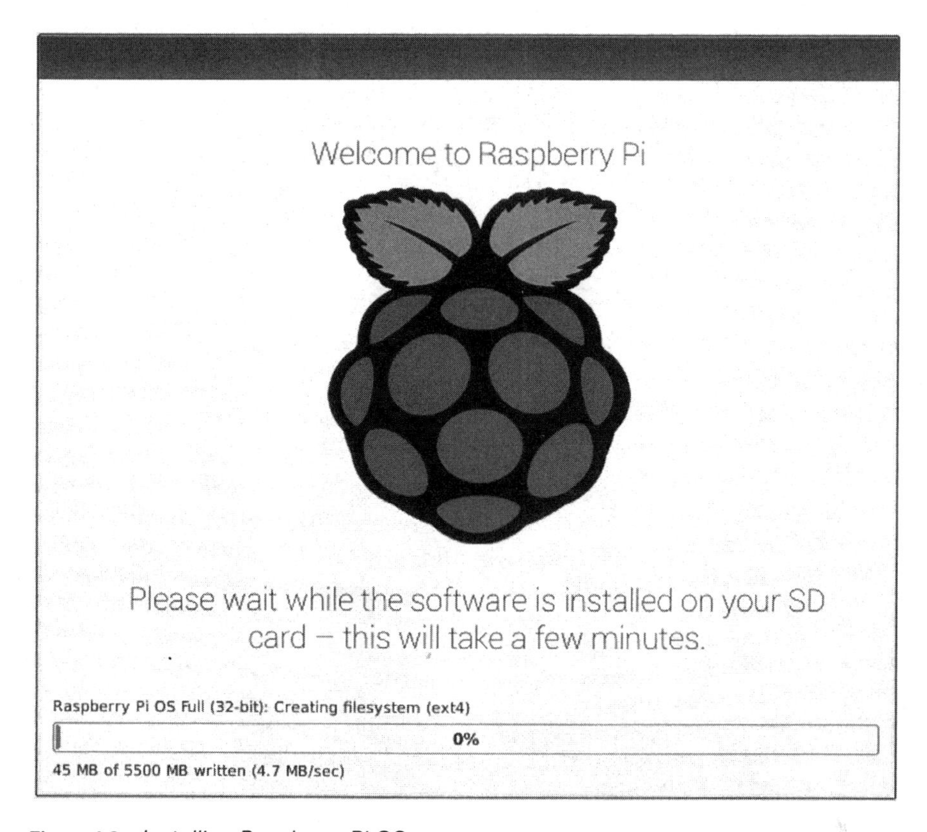

Figure 1-9 *Installing Raspberry Pi OS.*

Figure 1-10 *Selecting a Wi-Fi network.*

Summary

Now that we have set up our Raspberry Pi and it is ready to use, we can start exploring some of its features and get a grip on the basics of Linux.

2

Getting Started

The Raspberry Pi uses a distribution of Linux called Raspberry Pi OS as its operating system. This chapter introduces Linux and shows you how to use the desktop and command line.

Linux

Linux is an open source operating system. This software has been written as a community project for those looking for an alternative to the duopoly of Microsoft Windows and Apple OS X. It is a fully featured operating system based on the same solid UNIX concepts that arose in the early days of computing. It has a loyal and helpful following and has matured into an operating system that is powerful and easy to use.

Although the operating system is called Linux, various Linux distributions (or *distros*) have been produced. These involve the same basic operating system, but are packaged with different bundles of applications or different windowing systems. Although many distros are available, the one recommended by the Raspberry Pi Foundation is called Raspberry Pi OS.

If you are only used to some flavor of Microsoft Windows, expect to experience some frustration as you get used to a new operating system. Things work a little differently in Linux. Almost anything you may want to change about Linux can be changed. The system is open and completely under your control. However, as they say in *Spiderman,* with great power comes great responsibility. This means that if you are not careful, you could end up breaking your operating system.

The Desktop

At the end of Chapter 1, we had just booted up our Raspberry Pi, logged in, and started up the windowing system. Figure 2-1 serves to remind you of what the Raspberry Pi desktop looks like.

If you are a user of Windows or Mac computers, you will be familiar with the concept of a desktop as a folder within the file system that acts as a sort of background to everything you do on the computer.

Clicking the left-most icon on the bar at the top of the screen will show us some of the applications and tools installed on the Raspberry Pi (rather like the Start menu in Microsoft Windows). We are going to start with the File Manager, which can be found under the Accessories.

The File Manager is just like the File Explorer in Windows or the Finder on a Mac. It allows you to explore the file system, copy and move files, as well as launch files that are executable (applications).

When it starts, the File Manager shows you the contents of your home directory. You may remember that when you logged in, you gave your login name as pi. The root to your home directory will be /home/pi. Note that like Mac's OS X,

Figure 2-1 *Raspberry Pi desktop.*

Linux uses slash (/) characters to separate the parts of a directory name. Therefore, / is called the *root* directory and /home/ is a directory that contains other directories, one for each user. Our Raspberry Pi is just going to have one user (called pi), so this directory will only ever contain a directory called pi. The current directory is shown in the address bar at the top, and you can type directly into it to change the directory being viewed, or you can use the navigation bar at the side. The contents of the directory /home/pi include the Desktop and various other directories.

Double-clicking Desktop will open the Desktop directory, but this is not of much interest because it just contains the shortcuts on the left side of the desktop. If you open the bookshelf folder, you will see that it contains the single file containing a beginners guide to the Raspberry Pi, as shown in Figure 2-2.

You shouldn't often need to use any of the file system outside of your home directory. You should keep all documents, music files, and so on, housed within directories on your home folder or on an external USB flash drive.

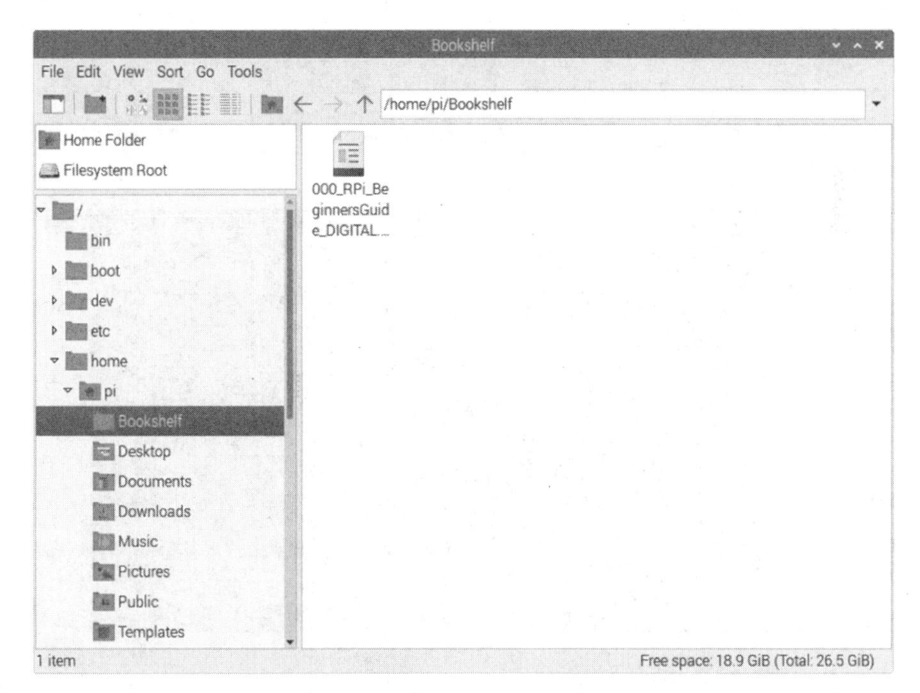

Figure 2-2 *The contents of bookshelf, as shown in File Manager.*

The Command Line

If you are a Windows or Mac user, you may have never used the command line. If you are a Linux user, on the other hand, you almost certainly will have done so. In fact, if you are a Linux user, then about now you will have realized that you probably don't need this chapter because it's all a bit basic for you.

Although it is possible to use a Linux system completely via the graphical interface, in general you will need to type commands into the command line. You do this to install new applications and to configure the Raspberry Pi.

To open a Terminal window, click on the Terminal icon at the top of the screen (looks like a monitor with a blank screen). This is a few icons to the right of the Raspberry P Menu (see Figure 2-3).

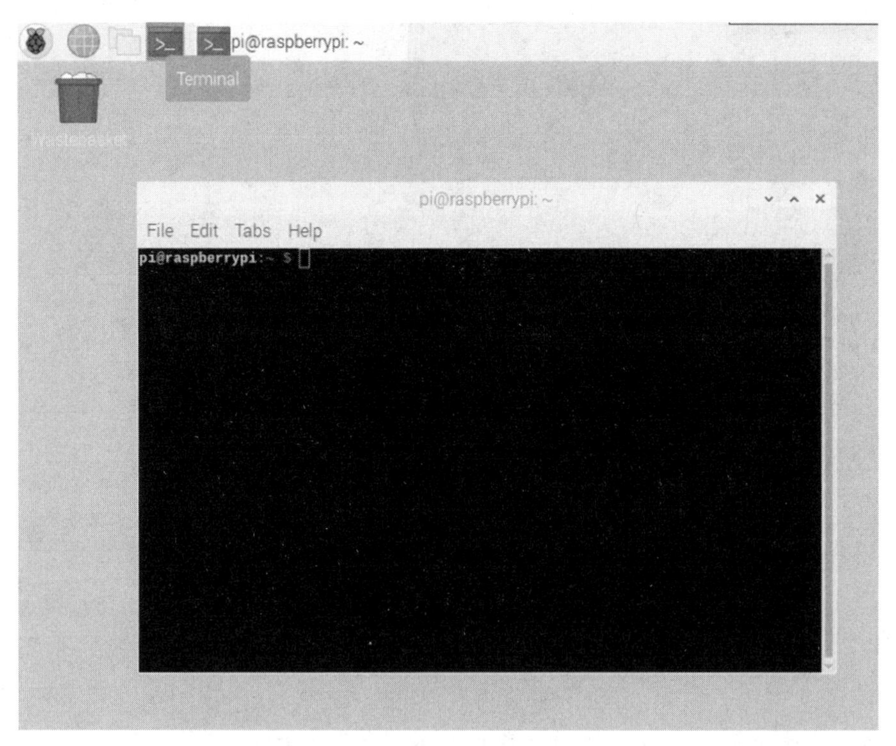

Figure 2-3 *The Terminal command line.*

Navigating with the Terminal

You will find yourself using three commands a lot when you are using the command line. The first command is pwd, which is short for *print working directory* and shows you which directory you are currently in. Therefore, after the $ sign in the terminal window, type **pwd** and press RETURN, as shown in Figure 2-4.

As you can see, we are currently in /home/pi. Rather than provide a screen shot for everything we are going to type into the terminal, I will use the convention that anything I want you to type will be prefixed with a $ sign, like this:

```
$ pwd
```

Anything you should see as a response will not have $ in front of it. Therefore, the whole process of running the pwd command would look something like this:

```
$ pwd
/home/pi
```

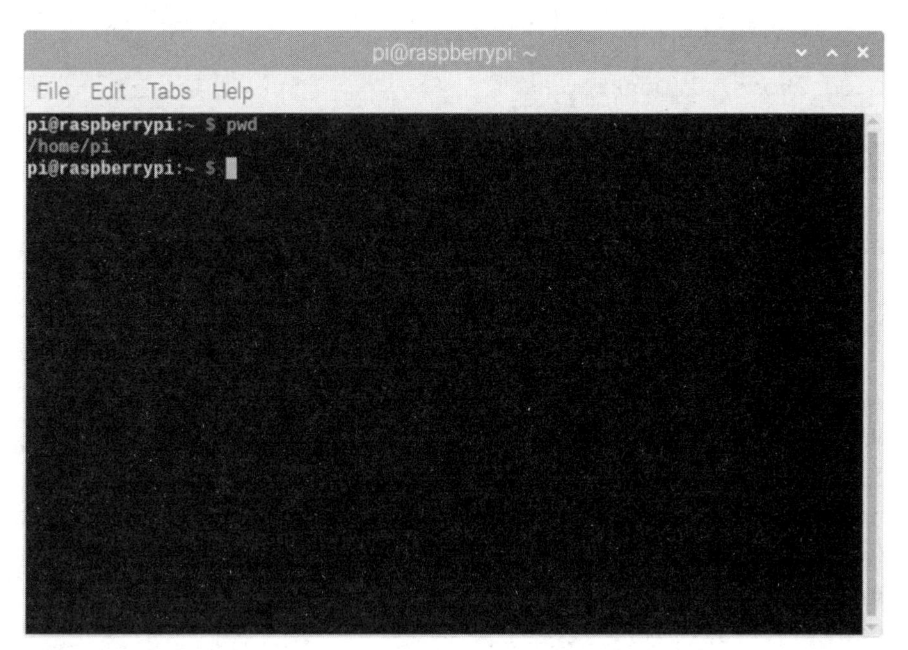

Figure 2-4 *The pwd command.*

The next common command we are going to discuss is `ls`, which is short for *list* and shows us a list of the files and directories within the working directory. Try the following:

```
$ ls
Bookshelf  Documents  Music     Public     Videos
Desktop    Downloads  Pictures  Templates
```

This tells us that the /home/pi folder contains the same set of folders that we saw using the File Manager in Figure 2-2.

The final command we are going to cover for navigating around is `cd` (which stands for *change directory*). This command changes the current working directory. It can change the directory relative either to the old working directory or to a completely different directory if you specify the whole directory, starting with /. So, for example, the following command will change the current working directory to /home/pi/Desktop:

```
$ pwd
/home/pi
$ cd Desktop
```

You could do the same thing by typing this:

```
$ cd /home/pi/Desktop
```

Note that when entering a directory or filename, you do not have to type all of it. Instead, at any time after you have typed some of the name, you can press the TAB key. If the filename is unique at that point, it will be automatically completed for you.

sudo

Another command that you will probably use a lot is `sudo` (for substitute-user do). This runs whatever command you type after it as if you were a super-user. You might be wondering why, as the sole user of this computer, you are not automatically a super-user. The answer is that, by default, your regular user account (username: pi, password: raspberry) does not have privileges that, say, allow you to go to some vital part of the operating system and start deleting files. Instead, to cause such mayhem, you have to prefix those commands with `sudo`. This just adds a bit of protection against accidents.

For the commands we have discussed so far, you will not need to prefix them with `sudo`. However, just for interest, try typing the following:

```
$ sudo ls
```

This will work the same way `ls` on its own works; you are still in the same working directory.

Applications

Take some time to look through the applications that are installed along with Raspberry Pi OS by clicking on the Raspberry icon in the top left of the screen and looking inside the various sub-menus. For example, Figure 2-5 shows the applications in the Office category. You can see here that you have word processor, spreadsheet, and presentation software akin to Microsoft Word, Excel, and PowerPoint.

Note that your menu may not have the same contents as Figure 2-5 as these often change between versions of Raspberry Pi OS.

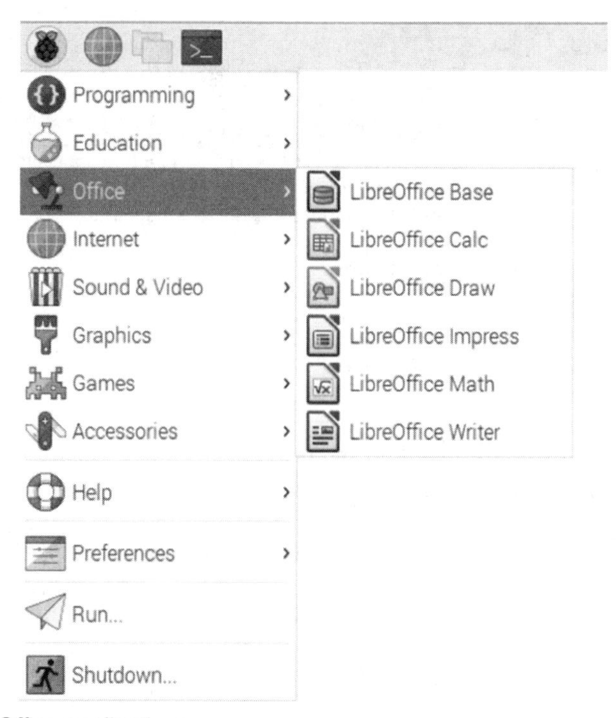

Figure 2-5 *Office applications.*

Other programs of note are:

- Mu within the Programming sub-menu
- The Chromium web browser within Internet
- A selection of games in Games
- Raspberry Pi Configuration within Preferences

Internet Resources

Aside from the business of programming the Raspberry Pi, you now have a functioning computer that you are probably keen to explore. To help you with this, many useful Internet sites are available where you can obtain advice and recommendations for getting the most out of your Raspberry Pi.

Table 2-1 lists some of the more useful sites relating to the Raspberry Pi. Your search engine will happily show you many more.

Site	Description
www.raspberrypi.org	The home page of the Raspberry Pi Foundation. Check out the forum and FAQs.
www.raspberrypi-spy.co.uk	A blog site with useful how-to posts.
http://elinux.org/RaspberryPiBoard	The main Raspberry Pi wiki. Lots of information about the Raspberry Pi, especially a useful list of verified peripherals (http://elinux.org/RPi_VerifiedPeripherals).

Table 2-1 *Internet Resources for the Raspberry Pi*

Summary

Now that we have everything set up and ready to go on our Raspberry Pi, it is time to start programming in Python.

3

Python Basics

The time has come to start creating some of our own programs for the Raspberry Pi. The language we are going to use is called Python. It has the great benefit that it is easy to learn while at the same time being powerful enough to create some interesting programs, including some simple games and programs that use graphics.

As with most things in life, it is necessary to learn to walk before you can run, and so we will begin with the basics of the Python language.

Okay, so a programming language is a language for writing computer programs in. But why do we have to use a special language anyway? Why couldn't we just use a human language? How does the computer use the things that we write in this language?

The reason why we don't use English or some other human language is that human languages are vague and ambiguous. Computer languages use English words and symbols, but in a very structured way.

Mu

The best way to learn a new language is to begin using it right away. So let's start up the program we are going to use to help us write Python. This program is called Mu, and you will find it in the Programming section of the Menu. You can see that Mu is asking us what we are going to be using Mu for. Select the option Python 3. Figure 3-1 shows what you will see when you first start Mu.

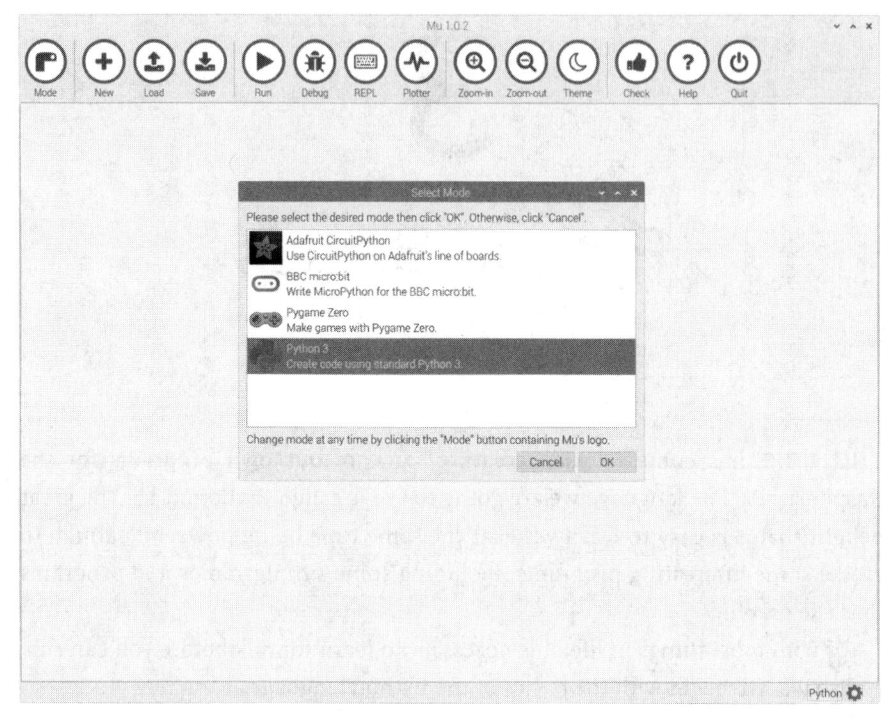

Figure 3-1 *Starting Mu for the first time.*

Python Versions

Python 3 was a major change over Python 2. This book is based on Python 3, but as you get further into Python you may find that some of the modules you want to use are not available for Python 3 and you need to revert to Python 2.

Python Shell

While you are learning Python, it's very helpful to be able to interactively type in Python commands to see what they do. To try this out, click on the REPL icon at the top of the Mu window and the window will split (Figure 3-2), opening up an area at the bottom of the screen where you can type commands. This area is called the REPL for Read-Evaluate-Print-Loop.

Rather like at the Linux command prompt, you can type in commands after the prompt (in this case, `IN [1]`) and the Python console will show you what it has done on the line below.

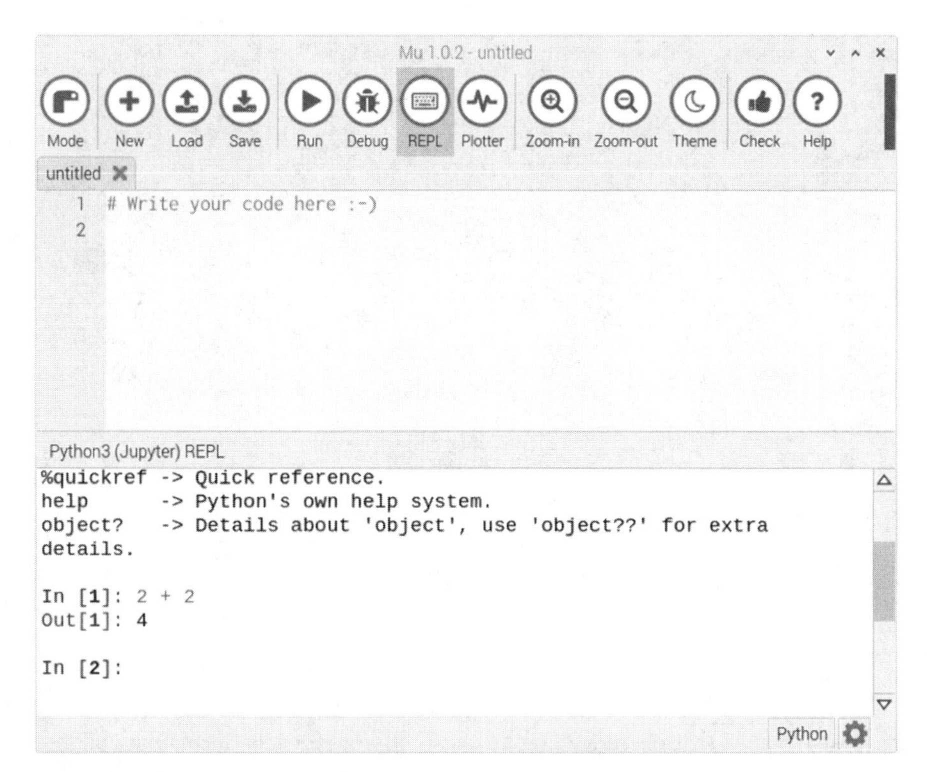

Figure 3-2 *Arithmetic in the REPL.*

Arithmetic is something that comes naturally to all programming languages, and Python is no exception. Therefore, type **2 + 2** after the prompt in the Python Shell and you should see the result (4) on the line below, as shown in Figure 3-2.

Editor

The REPL is a great place to experiment, but it is not the right place to write a program. Python programs are kept in files so that you do not have to retype them. A file may contain a long list of programming language commands, and when you want to run all the commands, what you actually do is run the file.

The menu bar at the top of Mu allows us to create a new file, but we don't need to use that because Mu has thoughtfully provided us with a new file ready to use. Note that at the moment there is some text in the file editor area that says: `'# Write your code here :-)'`. This is just a reminder for us as to where the code should go. You can delete this text now.

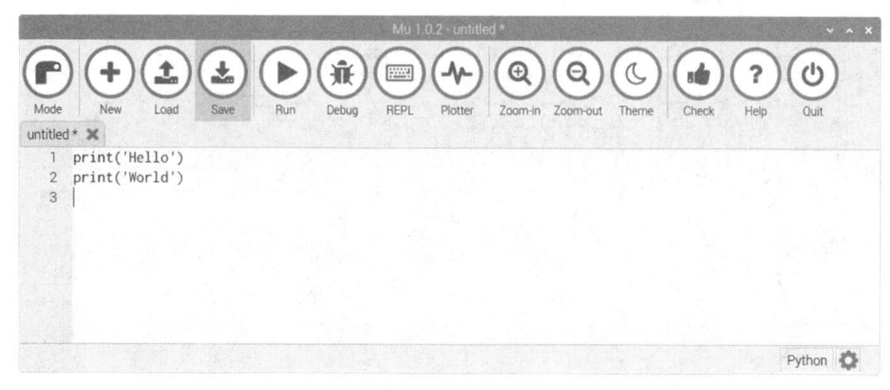

Figure 3-3 *The Mu editor.*

Click on the REPL icon again to close the REPL area of the window and then type the following two lines of code into the Mu editor window (Figure 3-3):

```
print('Hello')
print('World')
```

You will notice that the editor does not have the `In  [1]` prompt. This is because what we write here will not be executed immediately; instead, it will just be stored in a file until we decide to run it. If you wanted, you could use nano or some other text editor to write the file, but the Mu editor integrates nicely with Python. It also has some knowledge of the Python language and can thus serve as a memory aid when you are typing out programs.

The first time we run Mu, it will create a directory for us called '`mu_code`' in our home directory. This is where we are going to keep the Python programs that we write. So, click on the Save icon in Mu. This will open a dialog (Figure 3-4) so that you can give your program a name (**hello.py**) and choose a location to save it. Type **hello.py** in the File name field and then click Save.

Now that we have saved it, we can run the program and see what it does, click on the Run button. The results will appear at the bottom of the window (Figure 3-5). It is no great surprise that the program prints the two words *Hello* and *World,* each on its own line.

What you type in the REPL does not get saved anywhere; therefore, if you exit Mu and then start it up again, anything you typed in the REPL will be lost. However, because we saved our editor file (hello.py), we can load it at any time by clicking on the Load button.

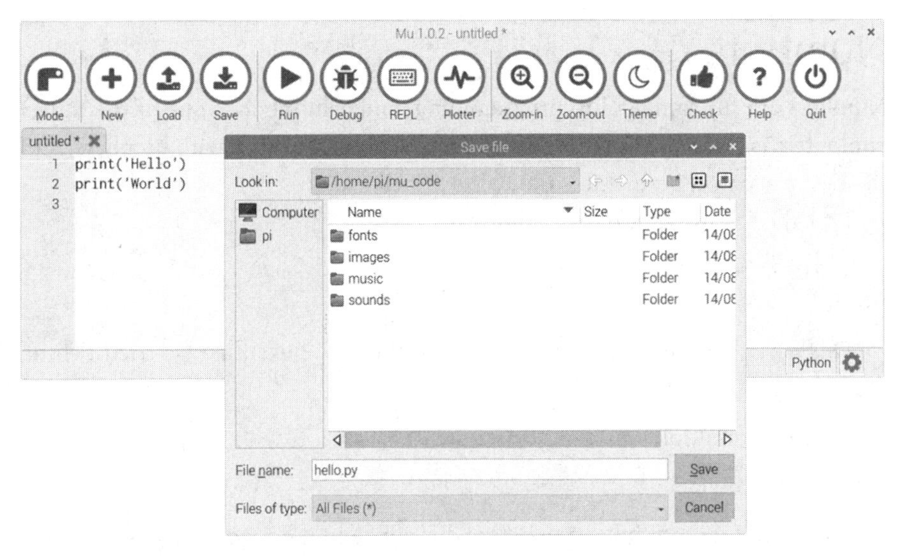

Figure 3-4 *Saving the hello.py program.*

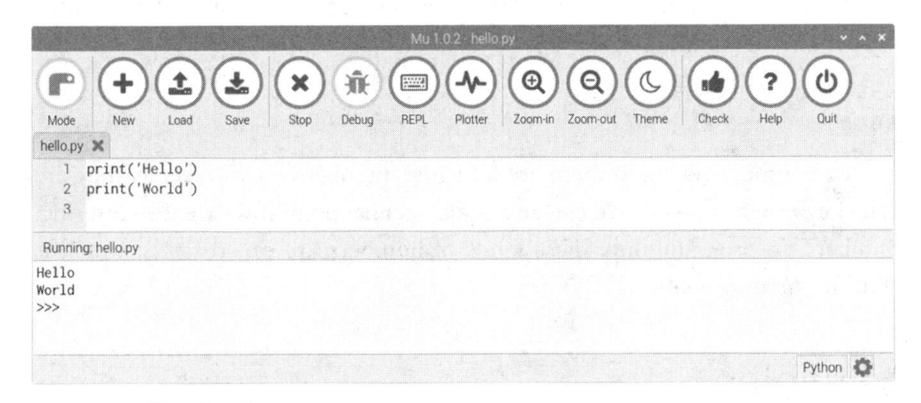

Figure 3-5 *Running the program.*

As you can see in Figure 3-5, after the two lines of hello world message have been printed, a prompt of >>> is shown. This is called the Python Shell and is just another version of the REPL. So, here you can type Python commands like the line '2 + 2' that we tried earlier.

NOTE *To save this book from becoming a series of screen dumps, from now on if I want you to type something in the REPL or Python Shell, I will precede it with >>>. The results will then appear on the lines below it.*

Numbers

Numbers are fundamental to programming, and arithmetic is one of the things computers are very good at. We will begin by experimenting with numbers, and the best place to experiment is the Python Shell.

Type the following into the Python Shell:

```
>>> 20 * 9 / 5 + 32
68.0
```

This isn't really advancing much beyond the 2 + 2 example we tried before. However, this example does tell us a few things:

- * means multiply.
- / means divide.
- Python does multiplication before division, and it does division before addition.

If you wanted to, you could add some parentheses to guarantee that everything happens in the right order, like this:

```
>>> (20 * 9 / 5) + 32
68.0
```

The numbers you have there are all whole numbers (or *integers* as they are called by programmers). We can also use a decimal point if we want to use such numbers. In programming, these kinds of numbers are called *floats,* which is short for *floating point.*

Variables

Sticking with the numbers theme for a moment, let's investigate variables. You can think of a variable as something that has a value. It is a bit like using letters as stand-ins for numbers in algebra. To begin, try entering the following:

```
>>> k = 9.0 / 5.0
```

The equals sign assigns a value to a variable. The variable must be on the left side and must be a single word (no spaces); however, it can be as long as you like and can contain numbers and the underscore character (_). Also, characters can be upper- and lowercase. Those are the rules for naming variables; however, there

Variable Name	Legal	Conventional
x	Yes	Yes
X	Yes	No
number_of_chickens	Yes	Yes
number of chickens	No	No
numberOfChickens	Yes	No
NumberOfChickens	Yes	No
2beOrNot2b	No	No
toBeOrNot2b	Yes	No

Table 3-1 *Naming Variables*

are also conventions. The difference is that if you break the rules, Python will complain, whereas if you break the conventions, other programmers may snort derisively and raise their eyebrows.

The conventions for variables are that they should start with a lowercase letter and should use an underscore between what in English would be words (for instance, `number_of_chickens`). The examples in Table 3-1 give you some idea of what is legal and what is conventional.

Many other languages use a different convention for variable names called bumpy-case or camel-case, where the words are separated by making the start of each word (except the first one) uppercase (for example, `numberOfChickens`). You will sometimes see this in Python example code. Ultimately, if the code is just for your own use, then how the variable is written does not really matter, but if your code is going to be read by others, it's a good idea to stick to the conventions.

By sticking to the naming conventions, it's easy for other Python programmers to understand your program.

If you do something Python doesn't like or understand, you will get an error message. Try entering the following:

```
>>> 2beOrNot2b = 1
SyntaxError: invalid syntax
```

This is an error because you are trying to define a variable that starts with a digit, which is not allowed.

A little while ago, we assigned a value to the variable `k`. We can see what value it has by just entering **k**, like so:

```
>>> k
1.8
```

Python has remembered the value of k, so we can now use it in other expressions. Going back to our original expression, we could enter the following:

```
>>> 20 * k + 32
68.0
```

For Loops

Arithmetic is all very well, but it does not make for a very exciting program. Therefore, in this section you will learn about *looping*, which means telling Python to perform a task a number of times rather than just once. In the following example, you will need to enter more than one line of Python. When you press RETURN and go to the second line, you will notice that Python is waiting. It has not immediately run what you have typed because it knows that you have not finished yet. The : character at the end of the line means that there is more to do.

These extra tasks must each appear on an indented line. To get this two-line program to actually run, press RETURN twice after the second line is entered.

```
>>> for x in range(1, 10):
        print(x)


1
2
3
4
5
6
7
8
9
>>>
```

This program has printed out the numbers between 1 and 9 rather than 1 and 10. The range command has an exclusive end point—that is, it doesn't include the last number in the range, but it does include the first.

You can check this out by just taking the range bit of the program and asking it to show its values as a list, like this:

```
>>> list(range(1, 10))
[1, 2, 3, 4, 5, 6, 7, 8, 9]
```

Some of the punctuation here needs a little explaining. The parentheses are used to contain what are called *parameters.* In this case, `range` has two parameters: `from` (`1`) and `to` (`10`), separated by a comma.

The `for in` command has two parts. After the word `for` there must be a variable name. This variable will be assigned a new value each time around the loop. Therefore, the first time it will be `1`, the next time `2`, and so on. After the word `in`, Python expects to see something that works out to be a list of items. In this case, this is a list of the numbers between 1 and 9.

The `print` command also takes an argument that displays it in the Python Shell. Each time around the loop, the next value of `x` will be printed out.

Simulating Dice

We'll now build on what you just learned about loops to write a program that simulates throwing a dice 10 times.

To do this, you will need to know how to generate a random number. So, first let's work out how to do that. If you didn't have this book, one way to find out how to generate a random number would be to type **random numbers python** into your search engine and look for fragments of code to type into the Python Shell. However, you do have this book, so here is what you need to write:

```
>>> import random
>>> random.randint(1,6)
2
```

Try entering the second line a few times, and you will see that you are getting different random numbers between 1 and 6. By the way, you can avoid having to type in a line you have typed before by pressing the up arrow on your keyboard. This will recall your previous commands in turn. When you have the one you want to execute, press the ENTER key.

The first line imports a library that tells Python how to generate numbers. You will learn much more about libraries later in this book, but for now you just need to know that we have to issue this command before we can start using the `randint` command that actually gives us a random number.

NOTE *I am being quite liberal with the use of the word* command *here. Strictly speaking, items such as* `randint` *are actually functions, not commands, but we will come to this later.*

Now that you can make a single random number, you need to combine this with your knowledge of loops to print off 10 random numbers at a time. This is getting beyond what can sensibly be typed into the Python Shell, so we will use the Mu editor.

You can either type in the examples from the text here or download all the Python examples used in the book from the book's web page simonmonk.org/prog-pi-ed3. Each programming example has a number. Thus, this program will be contained in the file 3_1_dice.py, which can be loaded into the Mu editor.

Installing the Example Programs

To copy all the example programs for this book onto your Raspberry Pi, make sure that it is connected to the Internet and then issue the following commands:

```
$ cd /home/pi/mu_code
$ git clone https://github.com/simonmonk/prog_pi_ed3.git
$ cd prog_pi_ed3
```

To see the full list of programs and other files needed in the book, list the contents of the directory using the `ls` command.

At this stage, it is worth typing in the examples to help the concepts sink in. Open up a new Mu editor tab (by clicking on the New button), type the following into it, and then save your work:

```
#3_1_dice
import random
for x in range(1, 11):
    random_number = random.randint(1, 6)
    print(random_number)
```

The first line begins with a # character. This indicates that the entire line is not program code at all, but just a comment to anyone looking at the program. Comments like this provide a useful way of adding extra information about a program into the program file, without interfering with the operation of the program. In other words, Python will ignore any line that starts with #.

Now, save the file giving it the **file name 3_1_dice.py** and then click on the Run button. The result should look something like Figure 3-6, where you can see the output below the editor window.

If you followed the instructions to install the example code, then you can open these examples by clicking on Load and then the prog_pi_ed3 folder (Figure 3-7), in which you will find all the example programs for the book.

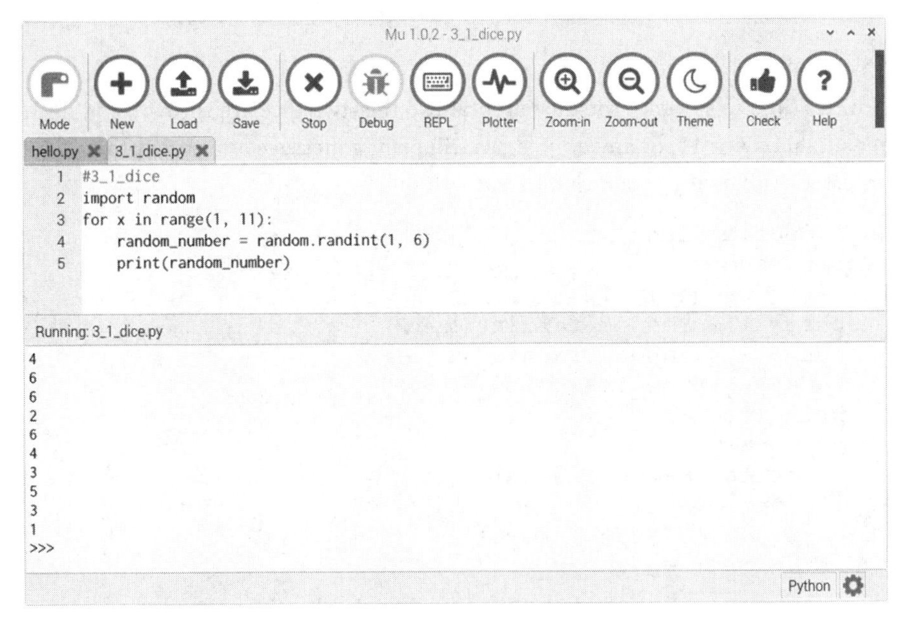

Figure 3-6 *The dice simulation.*

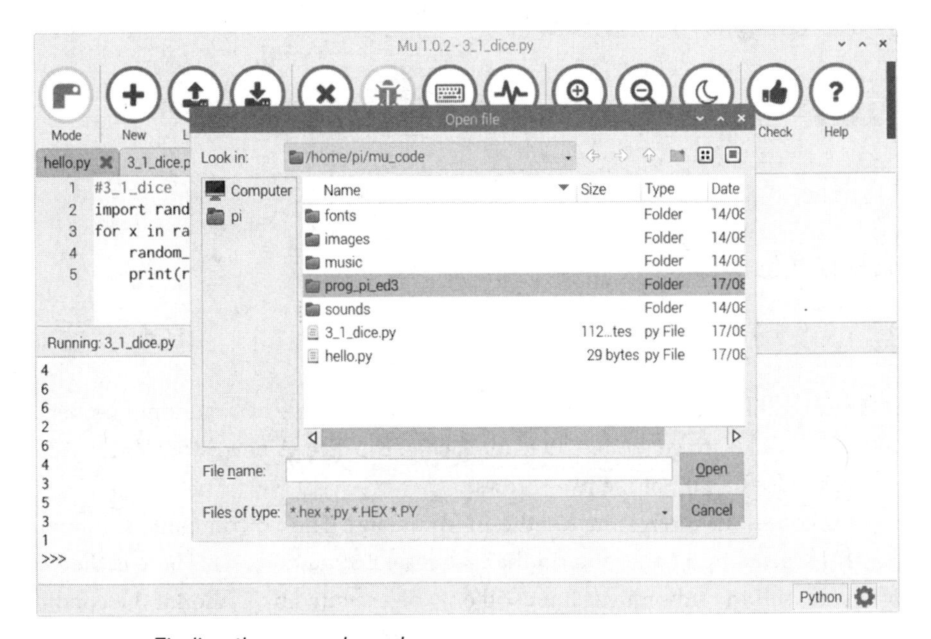

Figure 3-7 *Finding the example code.*

If

Now it's time to spice up the dice program so that two dice are thrown, and if we get a total of 7 or 11, or any double, we will print a message after the throw. Type or load the following program into Mu:

```
#3_2_double_dice
import random
for x in range(1, 11):
   throw_1 = random.randint(1, 6)
   throw_2 = random.randint(1, 6)
   total = throw_1 + throw_2
   print(total)
   if total == 7:
      print('Seven Thrown!')
   if total == 11:
      print('Eleven Thrown!')
   if throw_1 == throw_2:
      print('Double Thrown!')
```

When you run this program, you should see something like this:

```
6
7
Seven Thrown!
9
8
Double Thrown!
4
4
8
10
Double Thrown!
8
8
Double Thrown!
```

The first thing to notice about this program is that now two random numbers between 1 and 6 are generated. One for each of the dice. A new variable, `total`, is assigned to the sum of the two throws.

Next comes the interesting bit: the `if` command. The `if` command is immediately followed by a condition (in the first case, `total == 7`). There is then a colon (`:`), and the subsequent lines will only be executed by Python if the condition is true. At first sight, you might think there is a mistake in the condition because it uses == rather than =. The double equal sign is used when comparing items to see whether they are equal, whereas the single equal sign is used when assigning a value to a variable.

The second `if` is not tabbed in, so it will be executed regardless of whether the first `if` is true. This second `if` is just like the first, except that we are looking for a total of 11. The final `if` is a little different because it compares two variables (`throw_1` and `throw_2`) to see if they are the same, indicating that a double has been thrown.

Now, the next time you go to play *Monopoly* and find that the dice are missing, you know what to do: Just boot up your Raspberry Pi and write a little program.

Comparisons

To test to see whether two values are the same, we use ==. This is called a *comparison operator*. The comparison operators we can use are shown in Table 3-2.

You can do some experimenting with these comparison operators in the Python Shell. Here's an example:

```
>>> 10 > 9
True
```

In this case, we have basically said to Python, "Is 10 greater than 9?" Python has replied, "True." Now let's ask Python whether 10 is less than 9:

```
>>> 10 < 9
False
```

Being Logical

You cannot fault the logic. When Python tells us "True" or "False," it is not just displaying a message to us. `True` and `False` are special values called *logical values*.

Comparison	Description	Example
==	Equals	`total == 11`
!=	Not equals	`total != 11`
>	Greater than	`total > 10`
<	Less than	`total < 3`
>=	Greater than or equal to	`total >= 11`
<=	Less than or equal to	`total <= 2`

Table 3-2 *Comparison Operators*

Any condition we use with an `if` statement will be turned into a logical value by Python when it is deciding whether or not to perform the next line.

These logical values can be combined rather like the way you perform arithmetic operations like plus and minus. It does not make sense to add `True + True`, but it does make sense sometimes to say `True and True`.

As an example, if we wanted to display a message every time the total throw of our dice was between 5 and 9, we could write something like this:

```
if total >= 5 and total <= 9:
       print('not bad')
```

As well as `and`, we can use `or`. We can also use `not` to turn `True` into `False`, and vice versa, as shown here:

```
>>> not True
False
```

Thus, another way of saying the same thing would be to write the following:

```
if not (total < 5 or total > 9):
       print('not bad')
```

Exercise

Try incorporating the preceding test into the dice program. While you are at it, add two more `if` statements: one that prints "Good Throw!" if the throw is higher than 10 and one that prints "Unlucky!" if the throw is less than 4. Try your program out. If you get stuck, you can look at the solution in the file **03_03_double_dice_solution.py.**

Else

In the preceding example, you will see that some of the possible throws can be followed by more than one message. Any of the `if` lines could print an extra message if the condition is true. Sometimes you want a slightly different type of logic, so that if the condition is true, you do one thing and otherwise you do another. In Python, you use `else` to accomplish this:

```
>>> a = 7
>>> if a > 7:
       print('a is big')
else:
       print('a is small')
```

```
a is small
>>>
```

In this case, only one of the two messages will ever be printed.

Another variation on this is `elif`, which is short for *else if.* Thus, we could expand the previous example so that there are three mutually exclusive clauses, like this:

```
>>> a = 7
>>> if a > 9:
        print('a is very big')
elif a > 7:
        print('a is fairly big')
else:
        print('a is small')

a is small
>>>
```

While

Another command for looping is `while`, which works a little differently than `for`. The command `while` looks a bit like an `if` command in that it is immediately followed by a condition. In this case, the condition is for staying in the loop. In other words, the code inside the loop will be executed until the condition is no longer true. This means that you have to be careful to ensure that the condition will at some point be false; otherwise, the loop will continue forever and your program will appear to have hung.

To illustrate the use of `while`, the dice program has been modified so that it just keeps on rolling until a double 6 is rolled:

```
#3_4_double_dice_while
import random
throw_1 = random.randint(1, 6)
throw_2 = random.randint(1, 6)
while not (throw_1 == 6 and throw_2 == 6):
   total = throw_1 + throw_2
   print(total)
   throw_1 = random.randint(1, 6)
   throw_2 = random.randint(1, 6)
print('Double Six thrown!')
```

This program will work. Try it out. However, it is a little bigger than it should be. We are having to repeat the following lines twice—once before the loop starts and once inside the loop:

```
throw_1 = random.randint(1, 6)
throw_2 = random.randint(1, 6)
```

A well-known principle in programming is DRY (Don't Repeat Yourself). Although it's not a concern in a little program like this, as programs get more complex, you need to avoid the situation where the same code is used in more than one place, which makes the programs difficult to maintain.

We can use the command `break` to shorten the code and make it a bit "drier." When Python encounters the command `break`, it breaks out of the loop. Here is the program again, this time using `break`:

```
#3_5_double_dice_while_break
import random
while True:
    throw_1 = random.randint(1, 6)
    throw_2 = random.randint(1, 6)
    total = throw_1 + throw_2
    print(total)
    if throw_1 == 6 and throw_2 == 6:
        break
print('Double Six thrown!')
```

The condition for staying in the loop is permanently set to `True`. The loop will continue until it gets to `break`, which will only happen after throwing a double 6.

The Python Shell from the Terminal

Another way of running the Python shell is to use the Terminal. To do this, enter the command:

```
$ python3
```

Note that if you just use the command python rather than `python3`, the shell that starts will be for Python 2 not Python 3 and some of the examples in this book will not work.

After you enter the command `python3`, the Terminal will show the `>>>` Python prompt ready for you to enter Python commands (Figure 3-8).

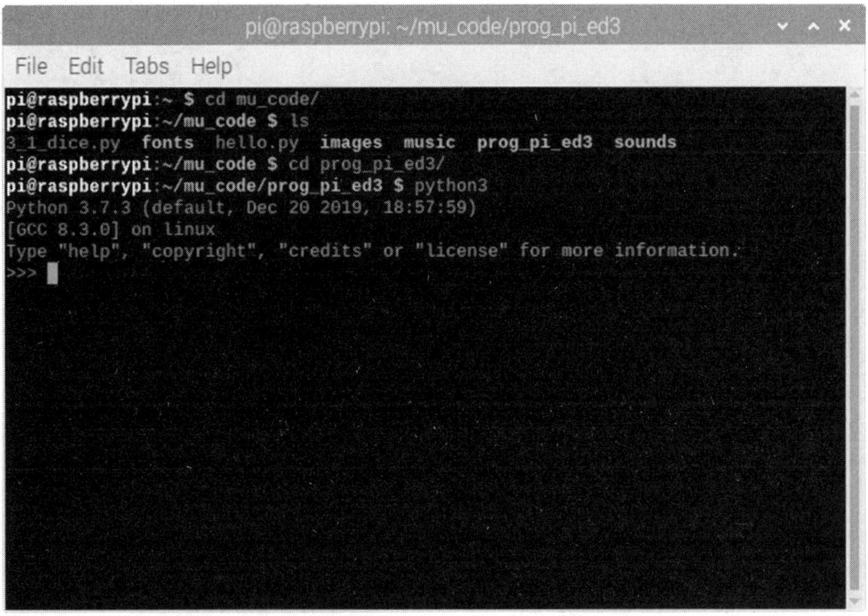

Figure 3-8 *Python in the Terminal.*

You can also run Python programs from the Terminal using the command `python3` followed by the name of the program you want to run. You can try this by entering the following commands:

```
$ cd /home/pi/mu_code/
$ python3 hello.py
```

Summary

You should now be happy to play with Mu and tryssse things out in the Python Shell. I strongly recommend that you try altering some of the examples from this chapter, changing the code and seeing how that affects what the programs do.

In the next chapter, we will move on past numbers to look at some of the other types of data you can work with in Python.

4

Strings, Lists, and Dictionaries

This chapter could have had "and Functions" added to its title, but the title was already long enough. In this chapter, you will first explore and play with the various ways of representing data and adding some structure to your programs in Python. You will then put everything you learned together into the simple game of Hangman, where you have to guess a word chosen at random by asking whether that word contains a particular letter.

The chapter ends with a reference section that tells you all you need to know about the most useful built-in functions for math, strings, lists, and dictionaries.

String Theory

No, this is not the Physics kind of String Theory. In programming, a *string* is a sequence of characters you use in your program. In Python, to make a variable that contains a string, you can just use the regular = operator to make the assignment, but rather than assigning the variable a number value, you assign it a string value by enclosing that value in single quotes, like this:

```
>>> book_name = 'Programming Raspberry Pi'
```

If you want to see the contents of a variable, you can do so either by entering just the variable name into the Python Shell or by using the `print` command, just as we did with variables that contain a number:

```
>>> book_name
'Programming Raspberry Pi'
>>> print(book_name)
Programming Raspberry Pi
```

There is a subtle difference between the results of each of these methods. If you just enter the variable name, Python puts single quotes around it so that you can tell it is a string. On the other hand, when you use `print`, Python just prints the value.

NOTE *You can also use double quotes to define a string, but the convention is to use single quotes unless you have a reason for using double quotes (for example, if the string you want to create has an apostrophe in it).*

You can find out how many characters a string has in it by doing this:

```
>>> len(book_name)
24
```

You can find the character at a particular place in the string like so:

```
>>> book_name[1]
'r'
```

Two things to notice here: first, the use of square brackets rather than the parentheses that are used for parameters and, second, that the positions start at 0 and not 1. To find the first letter of the string, you need to do the following:

```
>>> book_name[0]
'P'
```

If you put a number in that is too big for the length of the string, you will see this:

```
>>> book_name[100]
Traceback (most recent call last):
  File "<stdin>", line 1, in <module>
IndexError: string index out of range
>>>
```

This is an error, and it's Python's way of telling us that we have done something wrong. More specifically, the "string index out of range" part of the message tells us that we have tried to access something that we can't. In this case, that's element 100 of a string that is only 24 characters long.

You can chop lumps out of a big string into a smaller string, like this:

```
>>> book_name[0:11]
'Programming'
```

The first number within the brackets is the starting position for the string we want to chop out, and the second number is not, as you might expect, the position of the last character you want, but rather the last character plus 1.

As an experiment, try and chop out the word *raspberry* from the title. If you do not specify the second number, it will default to the end of the string:

```
>>> book_name[12:]
'Raspberry Pi'
```

Similarly, if you do not specify the first number, it defaults to 0.

Finally, you can also join strings together by using + operator. Here's an example:

```
>>> book_name + ' by Simon Monk'
'Programming Raspberry Pi by Simon Monk'
```

Lists

Earlier in the book when you were experimenting with numbers, a variable could only hold a single number. Sometimes, however, it is useful for a variable to hold a list of numbers or strings, or a mixture of both—or even a list of lists. Figure 4-1 will help you to visualize what is going on when a variable is a list.

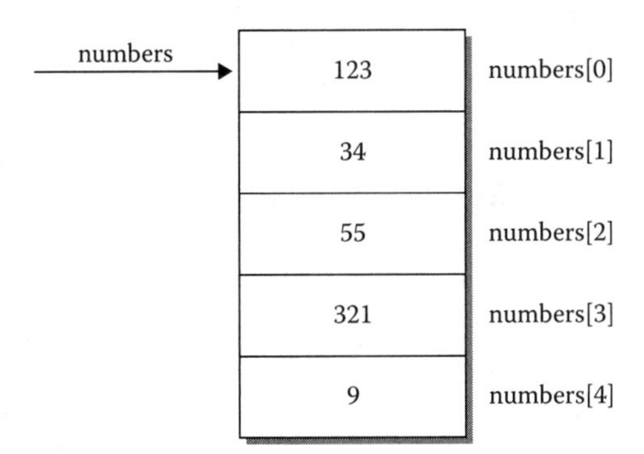

Figure 4-1 *A list.*

Lists behave rather like strings. After all, a string is a list of characters. The following example shows you how to make a list. Notice how `len` works on lists as well as strings:

```
>>> numbers = [123, 34, 55, 321, 9]
>>> len(numbers)
5
```

Square brackets are used to indicate a list, and just like with strings we can use square brackets to find an individual element of a list or to make a shorter list from a bigger one:

```
>>> numbers[0]
123
>>> numbers[1:3]
[34, 55]
```

What's more, you can use = to assign a new value to one of the items in the list, like this:

```
>>> numbers[0] = 1
>>> numbers
[1, 34, 55, 321, 9]
```

This changes the first element of the list (element 0) from 123 to just 1.

As with strings, you can join lists together using the + operator:

```
>>> more_numbers = [5, 66, 44]
>>> numbers + more_numbers
[1, 34, 55, 321, 9, 5, 66, 44]
```

If you want to sort the list, you can do this:

```
>>> numbers.sort()
>>> numbers
[1, 9, 34, 55, 321]
```

To remove an item from a list, you use the command pop, as shown next. If you do not specify an argument to pop, it will just remove the last element of the list and return it.

```
>>> numbers
[1, 9, 34, 55, 321]
>>> numbers.pop()
321
>>> numbers
[1, 9, 34, 55]
```

If you specify a number as the argument to pop, that is the position of the element to be removed. Here's an example:

```
>>> numbers
[1, 9, 34, 55]
>>> numbers.pop(1)
9
>>> numbers
[1, 34, 55]
```

As well as removing items from a list, you can also insert an item into the list at a particular position. The function `insert` takes two arguments. The first is the position before which to insert, and the second argument is the item to insert.

```
>>> numbers
[1, 34, 55]
>>> numbers.insert(1, 66)
>>> numbers
[1, 66, 34, 55]
```

When you want to find out how long a list is, you use `len(numbers)`, but when you want to sort the list or "pop" an element off the list, you put a dot after the variable containing the list and then issue the command, like this:

```
numbers.sort()
```

These two different styles are a result of something called *object orientation,* which we will discuss in the next chapter.

Lists can be made into quite complex structures that contain other lists and a mixture of different types—numbers, strings, and logical values. Figure 4-2 shows the list structure that results from the following line of code:

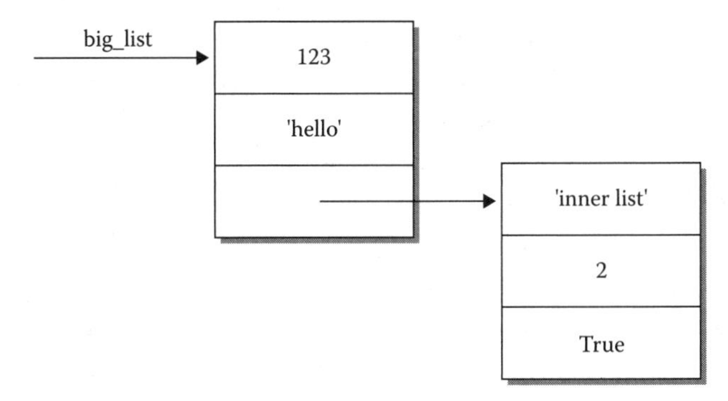

Figure 4-2 *A complex list.*

```
>>> big_list = [123, 'hello', ['inner list', 2, True]]
>>> big_list
[123, 'hello', ['inner list', 2, True]]
```

You can combine what you know about lists with for loops and write a short program that creates a list and then prints out each element of the list on a separate line:

```
#4_1_list_and_for
list = [1, 'one', 2, True]
for item in list:
    print(item)
```

Here's the output of this program:

```
1
one
2
True
```

Functions

When you are writing small programs like the ones we have been writing so far, they only really perform one function, so there is little need to break them up. It is fairly easy to see what they are trying to achieve. As programs get larger, however, things get more complicated and it becomes necessary to break up your programs into units called *functions.* When we get even further into programming, we will look at other ways still of structuring our programs using classes and modules.

Many of the things I have been referring to as *commands* are actually functions that are built into Python. Examples of this are range and print.

The biggest problem in software development of any sort is managing complexity. The best programmers write software that is easy to look at and understand and requires very little in the way of extra explanation. Functions are a key tool in creating easy-to-understand programs that can be changed without difficulty or risk of the whole thing falling into a crumpled mess.

A function is a little like a program within a program. We can use it to wrap up a sequence of commands we want to do. A function that we define can be called (used) from anywhere in our program and will contain its own variables and its own list of commands. When the commands have been run, we are returned to just after wherever it was in the code we called the function in the first place.

As an example, let's create a function that simply takes a string as an argument and adds the word *please* to the end of it. Load the following file—or even better, type it in to a new editor window—and then run it to see what happens:

```
#4_2_polite_function
def make_polite(sentence):
    polite_sentence = sentence + ' please'
    return polite_sentence

print(make_polite('Pass the salt'))
```

The function starts with the keyword `def`. This is followed by the name of the function, which follows the same naming conventions as variables. After that come the parameters inside parentheses and separated by commas if there are more than one. The first line must end with a colon.

Inside the function, we are using a new variable called `polite_sentence` that takes the parameter passed into the function and adds " please" to it (including the space at the start). This variable can only be used from inside the function.

The last line of the function is a `return` command. This specifies what value the function should give back to the code that called it. This is just like trigonometric functions such as `sin`, where you pass in an angle and get back a number. In this case, what is returned is the value in the variable `polite_sentence`.

To use the function, we just specify its name and supply it with the appropriate arguments. A return value is not mandatory, and some functions will just do something rather than calculate something. For example, we could write a rather pointless function that prints "Hello" a specified number of times:

```
#4_3_hello_n
def say_hello(n):
    for x in range(0, n):
        print('Hello')

say_hello(5)
```

This covers the basics of what we will need to do to write our game of Hangman. Although you'll need to learn some other things, we can come back to these later.

Hangman

Hangman is a word-guessing game, usually played with pen and paper. One player chooses a word and draws a dash for each letter of the word, and the other player has to guess the word. They guess a letter at a time. If the letter guessed is not in the word, they lose a life and part of the Hangman's scaffold is drawn. If the letter is in the word, all occurrences of the letter are shown by replacing the dashes with the letters.

We are going to let Python think of a word and we will have to guess what it is. Rather than draw a scaffold, Python is just going to tell us how many lives we have left.

Input in Python 2 and Python 3

This example is written in Python 3 and the finished Hangman example will cause an error if you try and run it as a Python 2 program, say by accidentally using the command `'python'` rather than `'python3'` from the command line.

The incompatibility arises because the '`input`' function works rather differently between the two versions of Python. In Python 3 the `input` command takes a parameter that is the prompt to the user as to what they are to type in as `input` to the program. When they have done this and hit ENTER then '`input`' will return whatever they typed as a string. Even if what you type is a number.

In Python 2, '`input`' tries to make sense of what you entered. So if you typed a number, it will return a number and if you type in something that starts with a letter, Python will assume that it's a variable and try and get its value. Generally, whatever you type is unlikely to be the name of a variable and you will get an error message.

The approach of Python 3 to reading `input` is much more consistent with other programming languages.

You can make the Hangman program work with Python 2 if you change every occurrence of '`input`' in the program with '`raw_input`'. The Python 2 '`raw_input`' function works just like '`input`' in Python 3.

You are going to start with how to give Python a list of words to choose from. This sounds like a job for a list of strings:

```
words = ['chicken', 'dog', 'cat', 'mouse', 'frog']
```

The next thing the program needs to do is to pick one of those words at random. We can write a function that does that and test it on its own:

```
#4_4_hangman_words
import random

words = ['chicken', 'dog', 'cat', 'mouse', 'frog']
```

```
def pick_a_word():
   return random.choice(words)

print(pick_a_word())
```

Run this program a few times to check that it is picking different words from the list. The 'choice' function from the 'random' module will, very helpfully, pick one of the items in the list at random.

This is a good start, but it needs to fit into the structure of the game. The next thing to do is to define a new variable called `lives_remaining`. This will be an integer that we can start off at 14 and decrease by 1 every time a wrong guess is made. This type of variable is called a *global* variable, because unlike variables defined in functions, we can access it from anywhere in the program.

As well as a new variable, we are also going to write a function called `play` that controls the game. We know what `play` should do, we just don't have all the details yet. Therefore, we can write the function `play` and make up other functions that it will call, such as `get_guess` and `process_guess`, as well as use the function `pick_a_word` we've just written. Here it is:

```
def play():
    word = pick_a_word()
    while True:
        guess = get_guess(word)
        if process_guess(guess, word):
            print('You win! Well Done!')
            break
        if lives_remaining == 0:
            print('You are Hung!')
            print('The word was: ' + word)
            break
```

A game of Hangman first involves picking a word. There is then a loop that continues until either the word is guessed (`process_guess` returns `True`) or `lives_remaining` has been reduced to zero. Each time around the loop, we ask the user for another guess.

We cannot run this at the moment because the functions `get_guess` and `process_guess` don't exist yet. However, we can write what are called *stubs* for them that will at least let us try out our `play` function. Stubs are just versions of functions that don't do much; they are stand-ins for when the full versions of the functions are written.

```
def get_guess(word):
    return 'a'
```

```
def process_guess(guess, word):
    global lives_remaining
    lives_remaining = lives_remaining - 1
    return False
```

The stub for `get_guess` just simulates the player always guessing the letter *a*, and the stub for `process_guess` always assumes that the player guessed wrong and, thus, decreases `lives_remaining` by 1 and returns `False` to indicate that they didn't win.

The stub for `process_guess` is a bit more complicated. The first line tells Python that the `lives_remaining` variable is the global variable of that name. Without that line, Python assumes that it is a new variable local to the function. The stub then reduces the lives remaining by 1 and returns `False` to indicate that the user has not won yet. Eventually, we will put in checks to see if the player has guessed all the letters or the whole word.

Open the file **04_05_hangman_play.py** and run it. You will get a result similar to this:

```
You are Hung!
The word was: dog
```

What happened here is that we have whizzed through all 14 guesses very quickly, and Python has told us what the word was and that we have lost.

All we need to do to complete the program is to replace the stub functions with real functions, starting with `get_guess`, shown here:

```
def get_guess(word):
    print_word_with_blanks(word)
    print('Lives Remaining: ' + str(lives_remaining))
    guess = input(' Guess a letter or whole word?')
    return guess
```

The first thing `get_guess` does is to tell the player the current state of their efforts at guessing (something like "`c--c--n`") using the function `print_word_with_blanks`. This is going to be another stub function for now. The player is then told how many lives they have left. Note that because we want to append a number (`lives_remaining`) after the string `Lives Remaining:`, the number variable must be converted into a string using the built-in `str` function.

The built-in function `input` prints the message in its parameter as a prompt and then returns anything that the user types.

Finally, the `get_guess` function returns whatever the user has typed.

The stub function `print_word_with_blanks` just reminds us that we have something else to write later:

```
def print_word_with_blanks(word):
    print('print_word_with_blanks: not done yet')
```

Open the file **04_06_hangman_get_guess.py** and run it. You will get a result similar to this:

```
not done yet
Lives Remaining: 14
 Guess a letter or whole word?x
print_word_with_blanks: not done yet
Lives Remaining: 13
 Guess a letter or whole word?y
print_word_with_blanks: not done yet
Lives Remaining: 12
 Guess a letter or whole word?
```

Enter guesses until all your lives are gone to verify that you get the "losing" message.

Next, we can create the proper version of `print_word_with_blanks`. This function needs to display something like "`c--c--n`," so it needs to know which letters the player has guessed and which they haven't. To do this, it uses a new global variable (this time a string) that contains all the guessed letters. Every time a letter is guessed, it gets added to this string:

```
guessed_letters = ''
```

Here is the function itself:

```
def print_word_with_blanks(word):
    display_word = ''
    for letter in word:
        if guessed_letters.find(letter) > -1:
            # letter found
            display_word = display_word + letter
        else:
            # letter not found
            display_word = display_word + '-'
    print display_word
```

This function starts with an empty string and then steps through each letter in the word. If the letter is one of the letters that the player has already guessed, it is added to display_word; otherwise, a hyphen (-) is added. The built-in function find is used to check whether the letter is in the guessed_letters. The find function returns -1 if the letter is not there; otherwise, it returns the position of the letter. All we really care about is whether or not it is there, so we just check that the result is greater than -1. Finally, the word is printed out.

Currently, every time process_guess is called, it doesn't do anything with the guess because it's still a stub. We can make it a bit less of a stub by having it add the guessed letter to guessed_letters, like so:

```
def process_guess(guess, word):
    global lives_remaining
    global guessed_letters
    lives_remaining = lives_remaining - 1
    guessed_letters = guessed_letters + guess
    return False
```

Open the file **04_07_hangman_print_word.py** and run it. You will get a result something like this:

```
-------
Lives Remaining: 14
 Guess a letter or whole word?c
c--c---
Lives Remaining: 13
 Guess a letter or whole word?h
ch-c---
Lives Remaining: 12
 Guess a letter or whole word?
```

It's starting to look like the proper game now. However, there is still the stub for process_guess to fill out. We will do that next:

```
def process_guess(guess, word):
    if len(guess) > 1:
        return whole_word_guess(guess, word)
    else:
        return single_letter_guess(guess, word)
```

When the player enters a guess, they have two choices: They can either enter a single-letter guess or attempt to guess the whole word. In this method, we just decide which type of guess it is and call either whole_word_guess or single_letter_guess. Because these functions are both pretty straightforward, we will implement them directly rather than as stubs:

```
def single_letter_guess(guess, word):
    global guessed_letters
    global lives_remaining
    if word.find(guess) == -1:
       # word guess was incorrect
       lives_remaining = lives_remaining - 1
    guessed_letters = guessed_letters + guess
    if all_letters_guessed(word):
       return True

def all_letters_guessed(word):
    for letter in word:
       if guessed_letters.find(letter) == -1:
           return False
    return True
```

The function whole_word_guess is actually easier than the single_letter_guess function:

```
def whole_word_guess(guess, word):
    global lives_remaining
    if guess.lower() == word.lower():
       return True
    else:
       lives_remaining = lives_remaining - 1
       return False
```

All we have to do is compare the guess and the actual word to see if they are the same when they are both converted to lowercase. We convert both to lowercase, so that it does not matter if the guessed word included some letters as capitals. We could just as easily convert both to uppercase and it would still work. If they are not the same, a life is lost. The function returns True if the guess was correct; otherwise, it returns False.

That's the complete program. Open up **04_08_hangman_full.py** in the Mu editor and run it. The full listing is shown here for convenience:

```
#04_08_hangman_full
import random

words = ['chicken', 'dog', 'cat', 'mouse', 'frog']
lives_remaining = 14
guessed_letters = ''
```

```
def play():
    word = pick_a_word()
    while True:
        guess = get_guess(word)
        if process_guess(guess, word):
            print('You win! Well Done!')
            break
        if lives_remaining == 0:
            print('You are Hung!')
            print('The word was: ' + word)
            break

def pick_a_word():
        return random.choice(words)

def get_guess(word):
    print_word_with_blanks(word)
    print('Lives Remaining: ' + str(lives_remaining))
    guess = input(' Guess a letter or whole word?')
    return guess

def print_word_with_blanks(word):
    display_word = ''
    for letter in word:
        if guessed_letters.find(letter) > -1:
          # letter found
          display_word = display_word + letter
        else:
          # letter not found
          display_word = display_word + '-'
    print(display_word)

def process_guess(guess, word):
    if len(guess) > 1:
        return whole_word_guess(guess, word)
    else:
        return single_letter_guess(guess, word)


def whole_word_guess(guess, word):
    global lives_remaining
    if guess == word:
        return True
    else:
        lives_remaining = lives_remaining - 1
        return False
```

```
def single_letter_guess(guess, word):
    global guessed_letters
    global lives_remaining
    if word.find(guess) == -1:
       # letter guess was incorrect
       lives_remaining = lives_remaining - 1
    guessed_letters = guessed_letters + guess
    if all_letters_guessed(word):
       return True
    return False

def all_letters_guessed(word):
    for letter in word:
       if guessed_letters.find(letter) == -1:
          return False
    return True

play()
```

The game as it stands has a few limitations. First, it is case sensitive, so you have to enter your guesses in lowercase, the same as the words in the `words` array. Second, if you accidentally type **aa** instead of **a** as a guess, it will treat this as a whole-word guess, even though it is too short to be the whole word. The game should probably spot this and only consider guesses the same length as the secret word to be whole-word guesses.

As an exercise, you might like to try and correct these problems. Hint: For the case-sensitivity problem, experiment with the built-in function `lower`. You can look at a corrected version in the file **04_08_hangman_full_solution.py.**

Dictionaries

Lists are great when you want to access your data starting at the beginning and working your way through, but they can be slow and inefficient when they get large and you have a lot of data to trawl through (for example, looking for a particular entry). It's a bit like having a book with no index or table of contents. To find what you want, you have to read through the whole thing.

Dictionaries, as you might guess, provide a more efficient means of accessing a data structure when you want to go straight to an item of interest. When you use a dictionary, you associate a value with a key. Whenever you want that value, you ask for it using the key. It's a little bit like how a variable name has a value associated with it; however, the difference is that with a dictionary, the keys and values are created while the program is running.

Let's look at an example:

```
>>> eggs_per_week = {'Penny': 7, 'Amy': 6, 'Bernadette': 0}
>>> eggs_per_week['Penny']
7
>>> eggs_per_week['Penny'] = 5
>>> eggs_per_week
{'Penny': 5, 'Amy': 6, 'Bernadette': 0}
>>>
```

This example is concerned with recording the number of eggs each of my chickens is currently laying. Associated with each chicken's name is a number of eggs per week. When we want to retrieve the value for one of the hens (let's say Penny), we use that name in square brackets instead of the index number that we would use with a list. We can use the same syntax in assignments to change one of the values.

For example, if Bernadette were to lay an egg, we could update our records by doing this:

```
eggs_per_week['Bernadette'] = 1
```

Note that although we have used a string as the key and a number as the value, the key could be a string, a number, or a tuple (see the next section), but the value could be anything, including a list or another dictionary.

Tuples

On the face of it, tuples look just like lists, but without the square brackets. Therefore, we can define and access a tuple like this:

```
>>> tuple = 1, 2, 3
>>> tuple
(1, 2, 3)
>>> tuple[0]
1
```

However, if we try to change an element of a tuple, we get an error message, like this one:

```
>>> tuple[0] = 6
Traceback (most recent call last):
  File "<stdin>", line 1, in <module>
TypeError: 'tuple' object does not support item assignment
```

The reason for this error message is that tuples are *immutable,* meaning that you cannot change them. Strings and numbers are also immutable. Although you can change a variable to refer to a different string, number, or tuple, you cannot change the number itself. On the other hand, if the variable references a list, you could alter that list by adding, removing, or changing elements in it.

So, if a tuple is just a list that you cannot do much with, you might be wondering why you would want to use one. The answer is, tuples provide a useful way of creating a temporary collection of items. Python lets you do a couple of neat tricks using tuples, as described in the next two subsections.

Multiple Assignment

To assign a value to a variable, you just use = operator, like this:

```
a = 1
```

Python also lets you do multiple assignments in a single line, like this:

```
>>> a, b, c = 1, 2, 3
>>> a
1
>>> b
2
>>> c
3
```

Multiple Return Values

Sometimes in a function, you want to return more than one value at a time. As an example, imagine a function that takes a list of numbers and returns the minimum and the maximum. Here is such an example:

```
#04_09_stats
def stats(numbers):
    numbers.sort()
    return (numbers[0], numbers[-1])

list = [5, 45, 12, 1, 78]
min, max = stats(list)
print(min)
print(max)
```

This method of finding the minimum and maximum is not terribly efficient, but it is a simple example. The list is sorted and then we take the first and last numbers. Note that `numbers[-1]` returns the last number because when you supply a negative index to an array or string, Python counts backward from the end of the list or string. Therefore, the position `-1` indicates the last element, `-2` the second to last, and so on.

Exceptions

Python uses exceptions to flag that something has gone wrong in your program. Errors can occur in any number of ways while your program is running. A common way we have already discussed is when you try to access an element of a list or string that is outside of the allowed range. Here's an example:

```
>>> list = [1, 2, 3, 4]
>>> list[4]
Traceback (most recent call last):
  File "<stdin>", line 1, in <module>
IndexError: list index out of range
```

If someone gets an error message like this while they are using your program, they will find it confusing to say the least. Therefore, Python provides a mechanism for intercepting such errors and allowing you to handle them in your own way:

```
try:
    list = [1, 2, 3, 4]
    list[4]
except IndexError:
    print('Oops')
```

We cover exceptions again in the next chapter, where you will learn about the hierarchy of the different types of error that can be caught.

Summary of Functions

This chapter was written to get you up to speed with the most important features of Python as quickly as possible. By necessity, we have glossed over a few things and left a few things out. Therefore, this section provides a reference of some of the key features and functions available for the main types we have discussed. Treat it as a resource you can refer back to as you progress though the book, and be sure to try out some of the functions to see how they work. There is no need to go through everything in this section—just know that it is here when you need it. Remember, the Python Shell is your friend.

For full details of everything in Python, refer to http://docs.python.org/py3k.

Numbers

Table 4-1 shows some of the functions you can use with numbers.

Function	Description	Example
`abs(x)`	Returns the absolute value (removes the - sign).	`>>>abs(-12.3)` `12.3`
`bin(x)`	Used to convert to binary string.	`>>> bin(23)` `'0b10111'`
`complex(r,i)`	Creates a complex number with real and imaginary components. Used in science and engineering.	`>>> complex(2,3)` `(2+3j)`
`hex(x)`	Used to convert to hexadecimal string.	`>>> hex(255)` `'0xff'`
`oct(x)`	Used to convert to octal string.	`>>> oct(9)` `'0o11'`
`round(x, n)`	Round x to n decimal places.	`>>> round(1.111111, 2)` `1.11`
`math.factorial(n)`	Factorial function (as in 4 × 3 × 2 × 1).	`>>> math.factorial(4)24`
`math.log(x)`	Natural logarithm.	`>>> math.log(10)` `2.302585092994046`
`math.pow(x, y)`	Raises x to the power of y (alternatively, use x ** y).	`>>> math.pow(2, 8)` `256.0`
`math.sqrt(x)`	Square root.	`>>> math.sqrt(16)` `4.0`
`math.sin, cos, tan, asin, acos, atan`	Trigonometry functions (radians).	`>>> math.sin(math.pi / 2)` `1.0`

Table 4-1 *Number Functions*

Strings

String constants can be enclosed either with single quotes (most common) or with double quotes. Double quotes are useful if you want to include single quotes in the string, like this:

```
s = "Its 3 o'clock"
```

On some occasions you'll want to include special characters such as end-of-lines and tabs into a string. To do this, you use what are called *escape characters,* which begin with a backslash (\) character. Here are the only ones you are likely to need:

- **`\t`** Tab character
- **`\n`** Newline character

Table 4-2 shows some of the functions you can use with strings.

Function	Description	Example
`s.capitalize()`	Capitalizes the first letter and makes the rest lowercase.	`>>> 'aBc'.capitalize()` `'Abc'`
`s.center(width)`	Pads the string with spaces, centering it. An optional extra parameter is used for the fill character.	`>>> 'abc'.center(10, '-')` `'---abc----'`
`s.endswith(str)`	Returns `True` if the end of the string matches.	`>>> 'abcdef'` `.endswith('def')` `True`
`s.find(str)`	Returns the position of a substring. Optional extra arguments for the start and end positions can be used to limit the search.	`>>> 'abcdef'.find('de')` `3`
`s.format(args)`	Formats a string using template markers using { }.	`>>> "Its {0} pm".format('12')` `"Its 12 pm"`
`s.isalnum()`	Returns `True` if all the characters in the string are letters or digits.	`>>> '123abc'.isalnum()` `True`
`s.isalpha()`	Returns `True` if all the characters are alphabetic.	`>>> '123abc'.isalpha()` `False`
`s.isspace()`	Returns `True` if the character is a space, tab, or other whitespace character.	`>>> ' \t'.isspace()` `True`
`s.ljust(width)`	Like `center()`, but left-justified.	`>>> 'abc'.ljust(10, '-')` `'abc-------'`
`s.lower()`	Converts a string into lowercase.	`>>> 'AbCdE'.lower()` `'abcde'`
`s.replace(old, new)`	Replaces all occurrences of `old` with `new`.	`>>> 'hello world'` `.replace('world', 'there')` `'hello there'`
`s.split()`	Returns a list of all the words in the string, separated by spaces. An optional parameter can be used to indicate a different splitting character. The end of line character (\n) is a popular choice.	`>>> 'abc def'.split()` `['abc', 'def']`
`s.splitlines()`	Splits the string on the newline character.	
`s.strip()`	Removes whitespace from both ends of the string.	`>>> '   a b   '.strip()` `'a b'`
`s.upper()`	Refer to `lower()`, earlier in this table.	

Table 4-2 *String Functions*

Lists

We have already looked at most of the features of lists. Table 4-3 summarizes these features.

Function	Description	Example

`del(a[i:j])`	Deletes elements from the array, from element `i` to element `j-1`.	`>>> a = ['a', 'b', 'c']`
		`>>> del(a[1:2])`
		`>>> a`
		`['a', 'c']`

`a.append(x)`	Appends an element to the end of the list.	`>>> a = ['a', 'b', 'c']`
		`>>> a.append('d')`
		`>>> a`
		`['a', 'b', 'c', 'd']`

`a.count(x)`	Counts the occurrences of a particular element.	`>>> a = ['a', 'b', 'a']`
		`>>> a.count('a')`
		`2`

`a.index(x)`	Returns the index position of the first occurrence of x in a. Optional parameters can be used for the start and end index.	`>>> a = ['a', 'b', 'c']`
		`>>> a.index('b')`
		`1`

`a.insert (i, x)`	Inserts x at position `i` in the list.	`>>> a = ['a', 'c']`
		`>>> a.insert(1, 'b')`
		`>>> a`
		`['a', 'b', 'c']`

`a.pop()`	Returns the last element of the list and removes it. An optional parameter lets you specify another index position for the removal.	`>>> ['a', 'b', 'c']`
		`>>> a.pop(1)`
		`'b'`
		`>>> a`
		`['a', 'c']`

`a.remove(x)`	Removes the element specified.	`>>> a = ['a', 'b', 'c']`
		`>>> a.remove('c')`
		`>>> a`
		`['a', 'b']`

`a.reverse()`	Reverses the list.	`>>> a = ['a', 'b', 'c']`
		`>>> a.reverse()`
		`>>> a`
		`['c', 'b', 'a']`

`a.sort()`	Sorts the list. Advanced options are available when sorting lists of objects. See the next chapter for details.	

Table 4-3 *List Functions*

Dictionaries

Table 4-4 details a few things about dictionaries that you should know.

Function	Description	Example

`len(d)`	Returns the number of items in the dictionary.	`>>> d = {'a':1, 'b':2}`
		`>>> len(d)`
		`2`

`del(d[key])`	Deletes an item from the dictionary.	`>>> d = {'a':1, 'b':2}`
		`>>> del(d['a'])`
		`>>> d`
		`{'b': 2}`

`key in d`	Returns `True` if the dictionary (d) contains the key.	`>>> d = {'a':1, 'b':2}`
		`>>> 'a' in d`
		`True`

`d.clear()`	Removes all items from the dictionary.	`>>> d = {'a':1, 'b':2}`
		`>>> d.clear()`
		`>>> d`
		`{}`

`get(key, default)`	Returns the value for the key, or default if the key is not there.	`>>> d = {'a':1, 'b':2}`
		`>>> d.get('c', 'c')`
		`'c'`

Table 4-4 *Dictionary Functions*

Type Conversions

We have already discussed the situation where we want to convert a number into a string so that we can append it to another string. Python contains some built-in functions for converting items of one type to another, as detailed in Table 4-5.

Function	Description	Example

`float(x)`	Converts x to a floating-point number.	`>>> float('12.34')`
		`12.34`
		`>>> float(12)`
		`12.0`

`int(x)`	Optional argument used to specify the number base.	`>>> int(12.34)`
		`12`
		`>>> int('FF', 16)`
		`255`

`list(x)`	Converts x to a list. This is also a handy way to get a list of dictionaries keys.	`>>> list('abc')`
		`['a', 'b', 'c']`
		`>>> d = {'a':1, 'b':2}`
		`>>> list(d)`
		`['a', 'b']`

Table 4-5 *Type Conversions*

Summary

Many things in Python you will discover gradually. Therefore, do not despair at the thought of learning all these commands. Doing so is really not necessary because you can always search for Python commands or look them up.

In the next chapter, we take the next step and see how Python manages object orientation.

5

Modules, Classes, and Methods

In this chapter, we discuss how to make and use our own modules, like the `random` module we used in Chapter 3. We also discuss how Python implements object orientation, which allows programs to be structured into classes, each responsible for its own behavior. This helps to keep a check on the complexity of our programs and generally makes them easier to manage. The main mechanisms for doing this are classes and methods. You have already used built-in classes and methods in earlier chapters without necessarily knowing it.

Modules

Most computer languages have a concept like modules that allows you to create a group of functions that are in a convenient form for others to use—or even for yourself to use on different projects.

Python does this grouping of functions in a very simple and elegant way. Essentially, any file with Python code in it can be thought of as a module with the same name as the file. However, before we get into writing our own modules, let's look at how we use the modules already installed with Python.

Using Modules

When we used the random module previously, we did something like this:

```
>>> import random
>>> random.randint(1, 6)
6
```

The first thing we do here is tell Python that we want to use the random module by using the import command. Somewhere in the Python installation is a file called random.py that contains the randint and choice functions as well as some other functions.

With so many modules available to us, there is a real danger that different modules might have functions with the same name. In such a case, how would Python know which one to use? Fortunately, we do not have to worry about this happening because we have imported the module, and none of the functions in the module are visible unless we prepend the module name and then a dot onto the front of the function name. Try omitting the module name, like this:

```
>>> import random
>>> randint(1, 6)
Traceback (most recent call last):
  File "<stdin>", line 1, in <module>
NameError: name 'randint' is not defined
```

Having to put the module name in front of every call to a function that's used a lot can get tedious. Fortunately, we can make this a little easier by adding to the import command as follows:

```
>>> import random as r
>>> r.randint(1,6)
2
```

This gives the module a local name within our program of just r rather than random, which saves us a bit of typing.

If you are certain a function you want to use from a library is not going to conflict with anything in your program, you can take things a stage further, as follows:

```
>>> from random import randint
>>> randint(1, 6)
5
```

To go even further, you can import everything from the module in one fell swoop. Unless you know exactly what is in the module, however, this is not normally a good idea, but you can do it. Here's how:

```
>>> from random import *
>>> randint(1, 6)
2
```

In this case, the asterisk (*) means "everything."

Useful Python Libraries

So far we have used the `random` module, but other modules are included in Python. These modules are often called Python's *standard library.* There are too many of these modules to list in full. However, you can always find a complete list of Python modules at http://docs.python.org/release/3.1.5/library/index.html. Here are some of the most useful modules you should take a look at:

- **`string`** String utilities
- **`datetime`** For manipulating dates and times
- **`math`** Math functions (sin, cos, and so on)
- **`pickle`** For saving and restoring data structures on file (see Chapter 6)
- **`urllib.request`** For reading web pages (see Chapter 6)
- **`guizero`** For creating graphical user interfaces (see Chapter 7)

Object Orientation

Object orientation has much in common with modules. It shares the same goals of trying to group related items together so that they are easy to maintain and find. As the name suggests, object orientation is about objects. We have been unobtrusively using objects already. A string is an object, for example. Thus, when we type

```
>>> 'abc'.upper()
```

We are telling the string `'abc'` that we want a copy of it, but in uppercase. In object-oriented terms, `abc` is an *instance* of the built-in class `str` and `upper` is a *method* on the class `str`.

We can actually find out the class of an object, as shown here (note double underscores before and after the word `class`):

```
>>> 'abc'.__class__
<class 'str'>
>>> [1].__class__
<class 'list'>
>>> 12.34.__class__
<class 'float'>
```

Defining Classes

That's enough of other people's classes; let's make some of our own. We are going to start by creating a class that does the job of converting measurements from one unit to another by multiplying a value by a scale factor.

We will give the class the catchy name `ScaleConverter`. Here is the listing for the whole class, plus a few lines of code to test it:

```
#05_01_converter
class ScaleConverter:
    def __init__(self, units_from, units_to, factor):
        self.units_from = units_from
        self.units_to = units_to
        self.factor = factor

    def description(self):
        return 'Convert ' + self.units_from + ' to ' + self.units_to

    def convert(self, value):
        return value * self.factor

c1 = ScaleConverter('inches', 'mm', 25)
print(c1.description())
print('converting 2 inches')
print(str(c1.convert(2)) + c1.units_to)
```

This requires some explanation. The first line is fairly obvious: It states that we are beginning the definition of a class called `ScaleConverter`. The colon (`:`) on the end indicates that all that follows is part of the class definition until we get back to an indent level of the left margin again.

Inside the ScaleConverter, we can see what look like three function definitions. These functions belong to the class; they cannot be used except via an instance of the class. These kinds of functions that belong to a class are called *methods.*

The first method, __init__, looks a bit strange—its name has two underscore characters on either side. When Python is creating a new instance of a class, it automatically calls the method __init__. The number of parameters that __init__ should have depends on how many parameters are supplied when an instance of the class is made. To unravel that, we need to look at this line at the end of the file:

```
c1 = ScaleConverter('inches', 'mm', 25)
```

This line creates a new instance of the ScaleConverter, specifying what the units being converted from and to are, as well as the scaling factor. The __init__ method must have all these parameters, but it must also have a parameter called self as the first parameter:

```
def __init__(self, units_from, units_to, factor):
```

The parameter self refers to the object itself. Now, looking at the body of the __init__ method, we see some assignments:

```
        self.units_from = units_from
        self.units_to = units_to
        self.factor = factor
```

Each of these assignments creates a variable that belongs to the object and has its initial value set from the parameters passed in to __init__.

To recap, when we create a new ScaleConverter by typing something like

```
c1 = ScaleConverter('inches', 'mm', 25)
```

Python creates a new instance of ScaleConverter and assigns the values 'inches', 'mm', and 25 to its three variables: self.units_from, self.units_to, and self.factor.

The term *encapsulation* is often used in discussions of classes. It is the job of a class to encapsulate everything to do with the class. That means storing data (like the three variables) and things that you might want to do with the data in the form of the description and convert methods.

The first of these (description) takes the information that the Converter knows about its units and creates a string that describes it. As with __init__, all methods must have a first parameter of `self`. The method will probably need it to access the data of the class to which it belongs.

Try it yourself by running program 05_01_converter.py and then typing the following in the Python Shell:

```
>>> silly_converter = ScaleConverter('apples', 'grapes', 74)
>>> silly_converter.description()
'Convert apples to grapes'
```

The `convert` method has two parameters: the mandatory `self` parameter and a parameter called `value`. The method simply returns the result of multiplying the value passed in by `self.factor`:

```
>>> silly_converter.convert(3)
222
```

Inheritance

The `ScaleConverter` class is okay for units of length and things like that; however, it would not work for something like converting temperature from degrees Celsius (C) to degrees Fahrenheit (F). The formula for this is F = C * 1.8 + 32. There is both a scale factor (1.8) and an offset (32).

Let's create a class called `ScaleAndOffsetConverter` that is just like `ScaleConverter`, but with a `factor` as well as an `offset`. One way to do this would simply be to copy the whole of the code for `ScaleConverter` and change it a bit by adding the extra variable. It might, in fact, look something like this:

```
#05_02_converter_offset_bad
class ScaleAndOffsetConverter:

    def __init__(self, units_from, units_to, factor, offset):
        self.units_from = units_from
        self.units_to = units_to
        self.factor = factor
        self.offset = offset

    def description(self):
        return 'Convert ' + self.units_from + ' to ' + self.units_to

    def convert(self, value):
        return value * self.factor + self.offset

c2 = ScaleAndOffsetConverter('C', 'F', 1.8, 32)
print(c2.description())
```

```
print('converting 20C')
print(str(c2.convert(20)) + c2.units_to)
```

Assuming we want both types of converters in the program we are writing, then this is a bad way of doing it. It's bad because we are repeating code. The `description` method is actually identical, and `__init__` is almost the same. A much better way is to use something called *inheritance.*

The idea behind inheritance in classes is that when you want a specialized version of a class that already exists, you inherit all the parent class's variables and methods and just add new ones or override the ones that are different. Figure 5-1 shows a class diagram for the two classes, indicating how `ScaleAndOffsetConverter` inherits from `ScaleConverter`, adds a new variable (`offset`), and overrides the method `convert` (because it will work a bit differently).

Here is the class definition for `ScaleAndOffsetConverter` using inheritance:

```
class ScaleAndOffsetConverter(ScaleConverter):

    def __init__(self, units_from, units_to, factor, offset):
        ScaleConverter.__init__(self, units_from, units_to, factor)
        self.offset = offset

    def convert(self, value):
        return value * self.factor + self.offset
```

The first thing to notice is that the class definition for `ScaleAndOffsetConverter` has `ScaleConverter` in parentheses immediately after it. That is how you specify the parent class for a class.

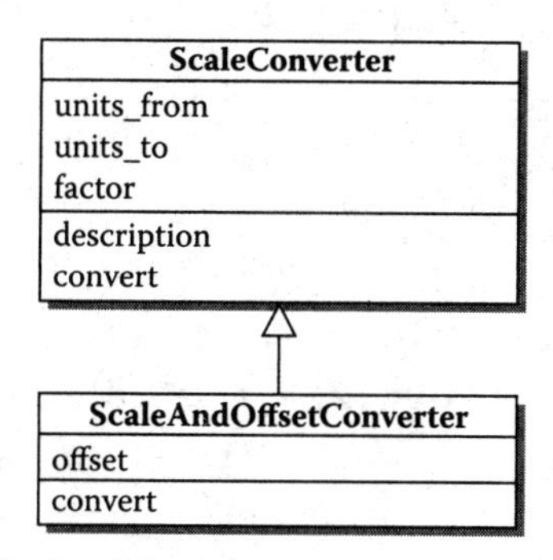

Figure 5-1 *An example of using inheritance.*

The `__init__` method for the new "subclass" of ScaleConverter first invokes the `__init__` method of ScaleConverter before defining the new variable offset. The convert method will override the convert method in the parent class because we need to add on the offset for this kind of converter. You can run and experiment with the two classes together by running 05_03_converters_final.py:

```
>>> c1 = ScaleConverter('inches', 'mm', 25)
>>> print(c1.description())
Convert inches to mm
>>> print('converting 2 inches')
converting 2 inches
>>> print(str(c1.convert(2)) + c1.units_to)
50mm
>>> c2 = ScaleAndOffsetConverter('C', 'F', 1.8, 32)
>>> print(c2.description())
Convert C to F
>>> print('converting 20C')
converting 20C
>>> print(str(c2.convert(20)) + c2.units_to)
68.0F
```

It's a simple matter to convert these two classes into a module that we can use in other programs. In fact, we will use this module in Chapter 7, where we attach a graphical user interface to it.

To turn this file into a module, we should first take the test code off the end of it and then give the file a more sensible name. Let's call it converters.py. You will find this file in the downloads for this book. The module must be in the same directory as any program that wants to use it.

To use the module now, just do this:

```
>>> import converters
>>> c1 = converters.ScaleConverter('inches', 'mm', 25)
>>> print(c1.description())
Convert inches to mm
>>> print('converting 2 inches')
converting 2 inches
>>> print(str(c1.convert(2)) + c1.units_to)
50mm
```

Summary

Lots of modules are available for Python, and some are specifically for the Raspberry Pi, such as the `gpiozero` library for controlling the GPIO pins. As you work through this book, you will encounter various modules. You will also find that as the programs you write get more complex, the benefits of an object-oriented approach to designing and coding your projects will keep everything more manageable.

In the next chapter, we look at using files and the Internet.

6

Files and the Internet

Python makes it easy for your programs to use files and connect to the Internet. You can read data from files, write data to files, and fetch content from the Internet. You can even check for new mail and tweet—all from your program.

Files

When you run a Python program, any values you have in variables will be lost. Files provide a means of making data more permanent.

Reading Files

Python makes reading the contents of a file extremely easy. As an example, we can convert the Hangman program from Chapter 4 to read the list of words from a file rather than have them fixed in the program.

First of all, start a new file in Mu and put some words in it, one per line. Then save the file with the name **hangman_words.txt** in the directory /home/pi/mu_code/prog_pi_ed3. Note that in the Save dialog you will have to change the file type to .txt (see Figure 6-1).

Before we modify the Hangman program itself, we can just experiment with reading the file in the Python console. Enter the following into the REPL:

```
>>> f = open('prog_pi_ed3/hangman_words.txt')
```

Note that REPL in Mu has a current directory of /home/mu/mu_code, so the directory of wherever you saved the file must be included in the `'open'` command.

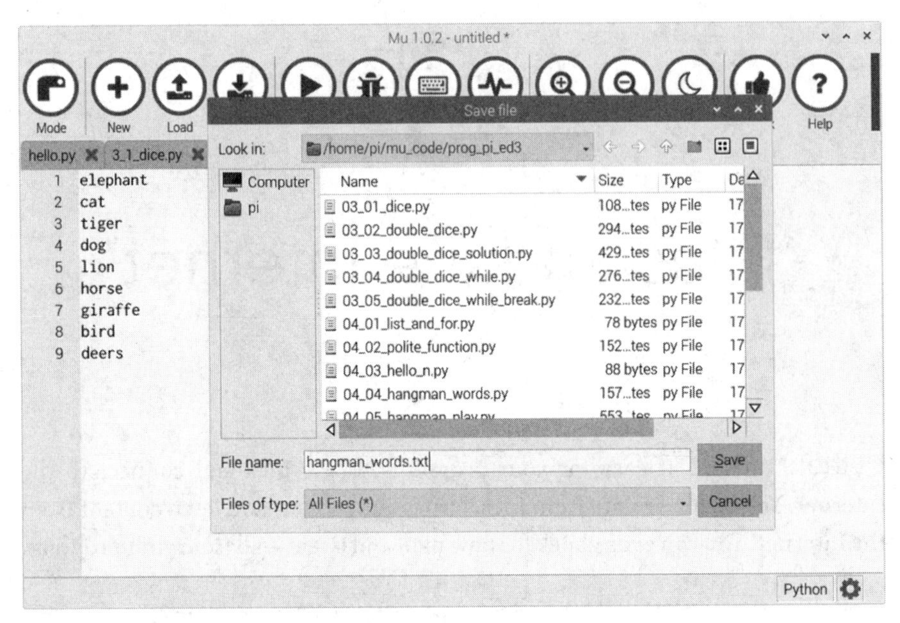

Figure 6-1 *Creating a text file in Mu.*

Next enter the following into the Python console:

```
>>> words = f.read()
>>> words
'elephant\ncat\ntiger\ndog\nlion\nhorse\ngiraffe\nbird\
ndeer\n'
>>> words.splitlines()
['elephant', 'cat', 'tiger', 'dog', 'lion', 'horse', 'gi-
raffe'
, 'bird', 'deer']
>>>
```

I told you it was easy! All we need to do to add this file to the Hangman program is replace the line

```
words = ['chicken', 'dog', 'cat', 'mouse', 'frog']
```

with the following lines:

```
f = open('prog_pi_ed3/hangman_words.txt')
words = f.read().splitlines()
f.close()
```

The line `f.close()` has been added. You should always call the `close` command when you are done with a file to free up operating system resources. Leaving a file open can lead to problems.

The full program is contained in the file **06_01_hangman_file.py,** and a suitable list of animal names can be found in the file hangman_words.txt. This program does nothing to check that the file exists before trying to read it. So, if the file isn't there, we get an error that looks something like this:

```
Traceback (most recent call last):
  File "06_01_hangman_file.py", line 4, in <module>
    f = open('hangman_words.txt')
IOError: [Errno 2] No such file or directory: 'hangman_words.txt'
```

To make this a bit more user friendly, the file-reading code needs to be inside a `try` command, like this:

```
try:
    f = open('prog_pi_ed3/hangman_words.txt')
    words = f.read().splitlines()
    f.close()
except IOError:
    print("Cannot find file 'prog_pi_ed3/hangman_words.txt'")
    exit()
```

Python will try to open the file, but because the file is missing it will not be able to. Therefore, the `except` part of the program will apply, and the more friendly message will be displayed. Because we cannot do anything without a list of words to guess, there is no point in continuing, so the `exit` command is used to quit.

In writing the error message, we have repeated the name of the file. Sticking strictly to the Don't Repeat Yourself (DRY) principle, the filename should be put in a variable, as shown next. That way, if we decide to use a different file, we only have to change the code in one place.

```
words_file = 'prog_pi_ed3/hangman_words.txt'
try:
        f = open(words_file)
        words = f.read().splitlines()
        f.close()
except IOError:
        print("Cannot find file: " + words_file)
        exit()
```

A modified version of Hangman with this code in it can be found in the file **06_02_hangman_file_try.py.**

Reading Big Files

The way we did things in the previous section is fine for a small file containing some words. However, if we were reading a really huge file (say, several megabytes), then two things would happen. First, it would take a significant amount of time for Python to read all the data. Second, because all the data is read at once, at least as much memory as the file size would be used, and for truly enormous files, that might result in Python running out of memory.

If you find yourself in the situation where you are reading a big file, you need to think about how you are going to handle it. For example, if you were searching a file for a particular string, you could just read one line of the file at a time, like this:

```
#06_03_file_readline
words_file = 'hangman_words.txt'
try:
    f = open(words_file)
    line = f.readline()
    while line != '':
        if line == 'elephant\n':
            print('There is an elephant in the file')
            break
        line = f.readline()
    f.close()
except IOError:
    print("Cannot find file: " + words_file)
```

When the function `readline` gets to the last line of the file, it returns an empty string (`''`). Otherwise, it returns the contents of the line, including the end-of-line character (`\n`). If it reads a blank line that is actually just a gap between lines and not the end of the file, it will return just the end-of-line character (`\n`). By the program only reading one line at a time, the memory being used is only ever equivalent to one full line.

If the file is not broken into convenient lines, you can specify an argument in `read` that limits the number of characters read. For example, the following will just read the first 20 characters of a file:

```
>>> f = open('prog_pi_ed3/hangman_words.txt')
>>> f.read(20)
'elephant\ncat\ntiger\nd'
>>> f.close()
```

Writing Files

Writing files is almost as simple. When a file is opened, as well as specifying the name of the file to open, you can also specify the mode in which to open the file.

The mode is represented by a character, and if no mode is specified it is assumed to be `r` for `read`. The modes are as follows:

- **`r (read)`**.
- **`w (write)`** Replaces the contents of any existing file with that name.
- **`a (append)`** Appends anything to be written onto the end of an existing file.
- **`r+`** Opens the file for both reading and writing (not often used).

You can also add `'b'` after `'r'`, `'w'`, or `'a'` to indicate that the file contains binary data rather than readable text.

To write a file, you open it with a second parameter of `'w'`, `'a'`, or `'r+'`. Here's an example:

```
>>> f = open('test.txt', 'w')
>>> f.write('This file is not empty')
>>> f.close()
```

Try finding the file using the file manager just to check it's there. You will find it in /home/pi/mu_code.

The File System

Occasionally, you will need to do some file-system-type operations on files (moving them, copying them, and so on). Python uses Linux to perform these actions, but provides a nice Python-style way of doing them. Many of these functions are in the `shutil` (shell utility) package. There's a number of subtle variations on the basic copy and move features that deal with file permissions and metadata. In this section, we just deal with the basic operations. You can refer to the official Python documentation for any other functions (http://docs.python.org/release/3.1.5/library).

Here's how to copy a file:

```
>>> import shutil
>>> shutil.copy('test.txt', 'test_copy.txt')
```

To move a file, either to change its name or move it to a different directory:

```
shutil.move('test_copy.txt', 'test_dup.txt')
```

This works on directories as well as files. If you want to copy an entire folder—including all its contents and its content's contents—you can use the function copytree. The rather dangerous function rmtree, on the other hand, will recursively remove a directory and all its contents—exercise extreme caution with this one!

The nicest way of finding out what is in a directory is via *globbing.* The package glob allows you to create a list of files in a directory by specifying a wildcard (*). Here's an example:

```
>>> import glob
glob.glob('*.txt')
['hangman_words.txt', 'test.txt', 'test_dup.txt']
```

If you just want all the files in the folder, you could use this:

```
glob.glob('*')
```

Pickling

Pickling involves saving the contents of a variable to a file in such a way that the file can be later loaded to get the original value back. The most common reason for wanting to do this is to save data between runs of a program. As an example, we can create a complex list containing another list and various other data objects and then pickle it into a file called mylist.pickle, like so:

```
>>> mylist = ['a', 123, [4, 5, True]]
>>> mylist
['a', 123, [4, 5, True]]
>>> import pickle
>>> f = open('mylist.pickle', 'wb')
>>> pickle.dump(mylist, f)
>>> f.close()
```

The pickle file is binary (that's why you open the file in 'wb' mode) and unfortunately you can't view it in a text editor. To reconstruct a pickle file into an object, here is what you do:

```
>>> f = open('mylist.pickle', 'rb')
>>> other_array = pickle.load(f)
>>> f.close()
>>> other_array
['a', 123, [4, 5, True]]
```

JSON

JSON is a text format for representing data. It is used both in files and in web services using JSON as a common interchange format for data. Python makes it really easy to use JSON by creating lists and dictionaries from JSON text retrieved from a file or from the Internet.

As an example, use Mu to create a file called "books.json" in /home/pi/mu_files and put the following JSON text into it.

```
{"books": [
    {"title": "Programming Raspberry Pi", "price": 10.95},
    {"title": "The Raspberry Pi Cookbook", "price": 14.95}
]}
```

When it comes to JSON, think of a { and } as enclosing a Python dictionary. In this case the dictionary has a key of the string "books" and a value that is a list contained between [and]. There are two values in that list, both of which are themselves dictionaries containing information about books.

Each book dictionary has keys of "title" and "price." The title is a string enclosed within quotes and the price is a number (and so does not need quotes around it).

The program file **06_04_json_file.py** reads the file books.json and converts it into Python dictionaries and lists for us. Load **06_04_json_file.py** into Mu and run it. You should see something like Figure 6-2.

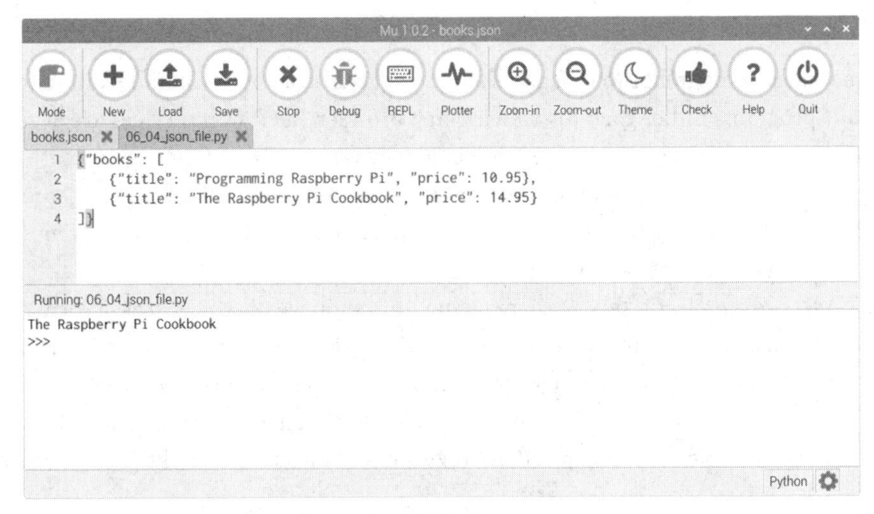

Figure 6-2 *Reading the contents of a JSON file.*

Let's have a look at the code for this.

```
#06_04_json_file
import json

f = open('books.json')
j = json.load(f)
f.close()

print(j['books'][1]['title'])
```

The first thing we do is import the json library. We then open the file books.json, and because json is in a text format, unlike a pickle file, we do not have to open it as a binary file.

The json method `'load'` reads the contents of the files and converts it into a dictionary. This is then assigned to the variable `'j'`. The files are then closed, as we have read what we need from it.

The final line of the program shows how you can use the [] notation to navigate into the structure of dictionaries and lists that have been created. This first uses `['books']` to retrieve the list of books and then uses `[1]` to select item 1 in the list (the second item) and then `['title']` selects the title of the second book.

In the next section you will learn how to use a web service to retrieve information about the weather from the Internet into your Python program.

Internet

As well as being full of web pages, the Internet is also a source of information for your programs rather than for viewing in a browser. Such web services are available for pretty much everything from currency rates and stock prices to weather forecasts. The software on the server for such web service is called an API (Application Programming Interface) and APIs generally use JSON as a way of communicating with your programs.

As an example of this, we are going to use a weather web service called weatherstack.com. In line with many such web services, you can use the web service for free as long as you don't make too many demands on it. However, you do normally have to register and make an account. weatherstack is no exception, so, to try out this example, you will need to register for an account at https://weatherstack.com/.

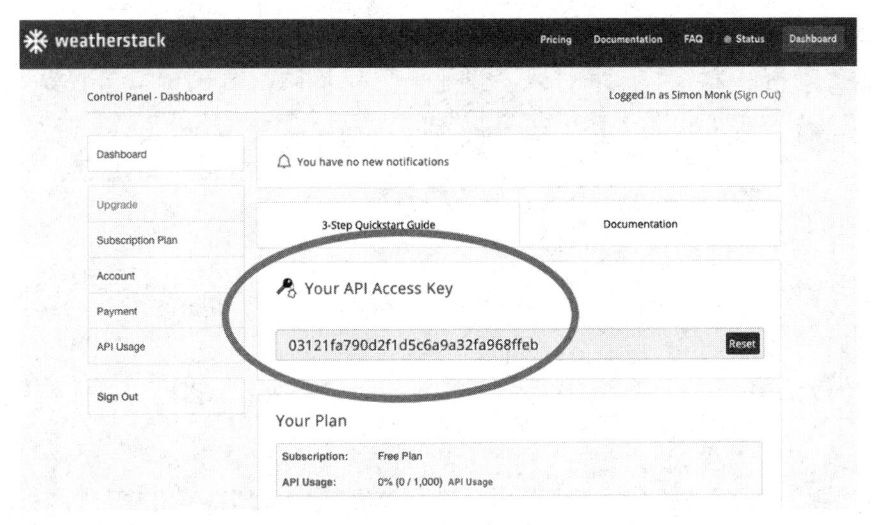

Figure 6-3 *Finding your API key in weatherstack.*

Not unreasonably, weatherstack likes to know who's using its API and so to access it from your programs you need to include your own personal key with each web request. Once you have registered, you can find your weatherstack key by clicking on the Dashboard button (Figure 6-3).

You will need to copy this key and paste it into the `'key = '` line of program 06_05_weather.py listed below and run the program.

```
#06_05_weather
import json
import urllib.parse, urllib.request

url = 'http://api.weatherstack.com/current'
city = urllib.parse.quote('San Francisco')
key = 'paste_your_key_here'

response = urllib.request.urlopen(url + '?access_key=' +
    key + '&query=' + city)
j = json.load(response)

print(j)
```

You should see output like that shown in Figure 6-4. You can change the city to your city.

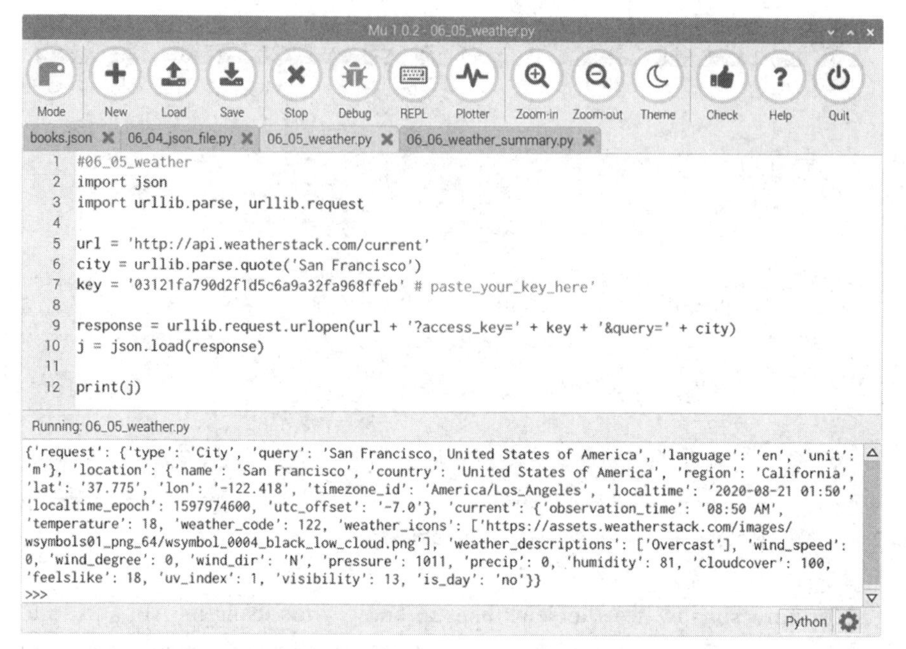

Figure 6-4 *Calling the weatherstack API.*

As you can see we have received a whole load of information back from the API that we can make use of in our programs.

The program uses the `urllib` library to perform the web request, contacting weatherstack's API, so this library has to be imported. Three variables are used, which will be assembled into the full URL to be sent to the API. They are the base URL, the API key (paste yours here), and the city. URLs cannot have spaces in them, but some city names (such as San Francisco) do. So, the method `urllib.parse.quote` is used to convert spaces into the %20 escape characters expected in a URL.

The method `urllib.request.urlopen` opens a connection to the API as if you were opening a file on your machine, and then json.load converts this to the JSON structure shown in Figure 6-4.

We probably don't want to see all this information, so we can navigate into the JSON and just pull out a summary by replacing the last line of the program with:

```
print(j['current']['weather_descriptions'][0])
```

Now, when you run the program, it will just display a message something like:

```
Partly cloudy
>>>
```

When you find a web service like this, that you want to use, find the documentation for the API so that you can see what kind of JSON you are going to need to deal with.

Summary

This chapter has given you the basics of how to use files and access web pages from Python. There is actually a lot more to Python and the Internet, including accessing e-mail and other Internet protocols. For more information on this, have a look at the Python documentation at http://docs.python.org/release/3.1.5/library/internet.html.

7

Graphical User Interfaces

Everything we have done so far has been text based. In fact, our Hangman game would not have looked out of place on a 1980s home computer. This chapter shows you how to create applications with a proper graphical user interface (GUI).

guizero

There are many Python libraries for creating GUIs, but perhaps the easiest to get started with is called guizero. guizero was created by Laura Sach and Martin O'Hanlan at the Raspberry Pi Foundation.

The guizero package is pre-installed on new versions of Raspberry Pi OS, so it's ready waiting for us to use. However, it's a good idea to make sure you have the latest version by running the following command in the terminal.

```
$ sudo pip3 install --upgrade guizero
```

Hello World

Tradition dictates that the first program you write with a new language or system should do something trivial, just to show it works! This usually means making the program display a message of "Hello World" (Figure 7-1). As you'll recall, we already did this for Python back in Chapter 3, so I'll make no apologies for starting with this program. Open it in Mu and then run it.

Figure 7-1 *Hello World in guizero.*

```
#07_01_hello.py

from guizero import App, Text
app = App()
Text(app, text="Hello World")
app.display()
```

We import two things from the *guizero* library: *App* represents the application window and *Text* represents a text label that we are going to use to hold our Hello World text. A new *App* is created and assigned to the variable *app*. We then make a *Text*, passing it *app* as its first parameter to that the text label will appear inside the application window. Finally we call `app.display()` to make the window visible.

You can just close the window by clicking on the x in the top right, or stop the program from Mu.

Temperature Converter

To get started with guizero, you'll gradually build up a simple application that provides a GUI for temperature conversion (see Figure 7-2). This application will use the `converter` module we created in Chapter 5 to do the calculation.

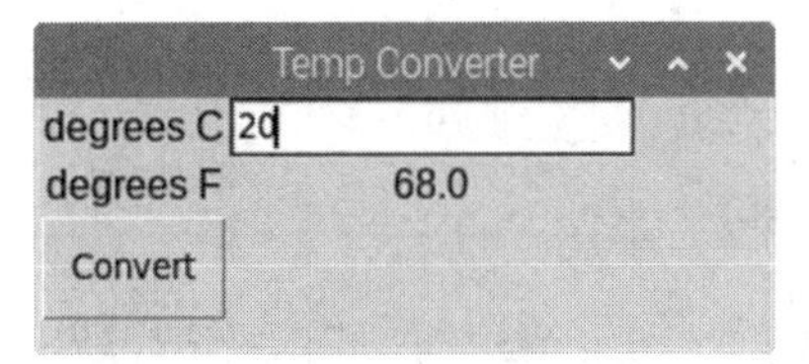

Figure 7-2 *The final temperature converter user interface.*

Our first step is to make a user interface that looks like Figure 7-2 but doesn't actually do anything:

```
#07_02_temp_gui.py

from guizero import *

app = App(title="Temp Converter", layout="grid",
          width=300, height=100)
Text(app, text="degrees C", grid=[0,0])
degCfield = TextBox(app, grid=[1,0], width="fill")

Text(app, text="degrees F", grid=[0,1]
degFfield = Text(app, grid=[1,1]

button = PushButton(app, text="Convert", grid=[0,2])

app.display()
```

We are going to need to import a few things from guizero, so we will just import everything (*) from it. We then create an App like we did for the Hello World application, but this time we are supplying some extra parameters when creating the App. These are as follows: a `title` which will appear in the window title area as well as the starting window size, specified as the `width` and `height` in pixels.

The other parameter (layout="grid") specifies that we are going to lay out the components on our window in a grid. These means that each component we add to the screen will have to specify its grid position. You can see this in the first Text label (degrees C) which specifies a grid position of `[0,0]` which means top left.

Figure 7-3 shows how we are going to lay out the components for this program in a grid.

By using a grid layout, we can line up the fields into columns and rows. The grid layout has three rows and two columns. The first row contains the label and

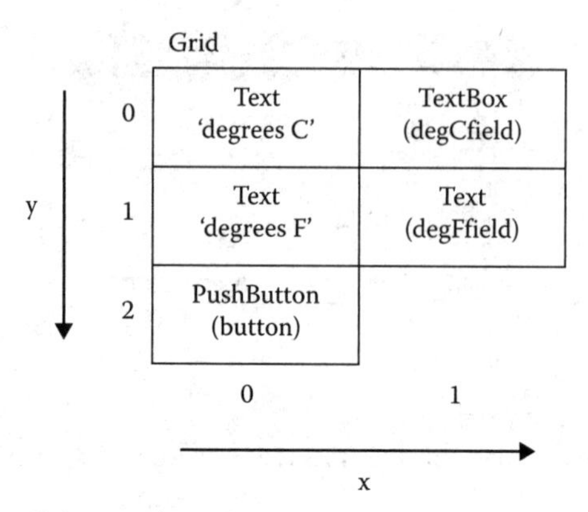

Figure 7-3 *Structure of the user interface.*

field to enter the temperature in degrees C. The second row has a label and field to show the result in Fahrenheit and the third row just contains the button.

When specifying the grid coordinates, you use an array of two elements. The first element is the x position (the column) and the second the y position (or row). The x positions go from left to right starting at 0 and the y positions (rows) start at 0 and begin at the top of the window and work down.

To make the `degCfield` larger, the attribute `width="fill"` will enlarge the control to fill the cell.

The example code above includes a button with the label of Convert, but if you try clicking on it, nothing will happen. Now that we have the user interface for the example, it's time to make it actually do the conversion. The final version of the program is listed below, with the additions marked in bold.

```
#07_03_temp_final.py

from guizero import *
from converters import ScaleAndOffsetConverter

c_to_f_conv = ScaleAndOffsetConverter('C', 'F', 1.8, 32)

def convert():
    c = float(degCfield.value)
    degFfield.value = str(c_to_f_conv.convert(c))

app = App(title="Temp Converter", layout="grid",
    width=300, height=100)
```

```
Text(app, text="degrees C", grid=[0,0])
degCfield = TextBox(app, grid=[1,0], width="fill")

Text(app, text="degrees F", grid=[0,1])
degFfield = Text(app, grid=[1,1])

button = PushButton(app, text="Convert", grid=[0,2],
command=convert)

app.display()
```

Although it's overkill for a little bit of arithmetic, we will use the `converters` module that we made in Chapter 5. From this module, we import the `ScaleAndOffsetConverter` and create an instance for converting degrees Centigrade to Fahrenheit.

The new function `convert` will be called when the Convert button is pressed. This reads whatever text has been typed into the `degCfield` and converts it from a string to a float. Note there is no checking to ensure that what has been typed into the field is a number, so typing non-numbers will cause an error. You may like to improve the program by adding error handling for this case. The value to be displayed in the `degFfield` is calculated using the converter and then converted into a string before being set to be the field's value.

The rest of the code is the same as the previous example except for the additional attribute `command` in the button definition. This links the `convert` function with the button.

Other GUI Widgets

In the temperature converter, we just used text fields (class TextBox) and labels (class Text). As you would expect, you can build lots of other user interface controls into your application. Figure 7-4 shows the main screen of a "kitchen sink" application that illustrates most of the controls you can use in guizero. This program is available as 07_04_kitchen_sink.py.

Here's the code for this program:

```
#07_04_kitchen_sink.py

from guizero import *

app = App(title="Kitchen Sink", layout="grid",
    width=400, height=400)
```

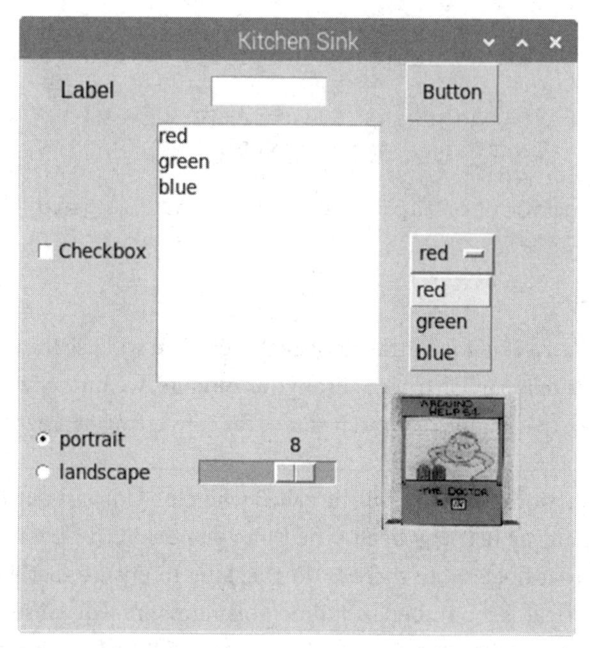

Figure 7-4 *A "Kitchen Sink" user interface.*

```
# Row 0
Text(app, text="Label", grid=[0,0])
TextBox(app, grid=[1,0])
PushButton(app, text="Button", grid=[2,0])

# Row 1
CheckBox(app, text="Checkbox", grid=[0,1])
ListBox(app, items=["red", "green", "blue"], grid=[1,1])
Combo(app, options=["red", "green", "blue"], grid=[2,1])

# Row 2
ButtonGroup(app, options=["portrait", "landscape"],
selected="portrait", grid=[0,2])
Slider(app, start=0, end=10, grid=[1,2])
Picture(app, image="prog_pi_ed3/test.png", width=100,
    height=100, grid=[2,2])

app.display()
```

This will get you started building your own interface, but you will very rapidly find that you need to start changing things like the alignment, font size, and other

Figure 7-5 *The Drawing control.*

aspects of the controls. You will find documentation on how to do all this in the official guizero documentation here: https://lawsie.github.io/guizero.

An interesting control provided by guizero is Drawing. This allows you to draw shapes and text in multiple colors. Figure 7-5 shows the kind of things you can do with Drawing.

Here's the listing for the code that created Figure 7-5:

```
#07_05_drawing.py

from guizero import *

app = App(width=400, height=200)
drawing = Drawing(app, width="fill", height="fill")
drawing.rectangle(20, 20, 300, 100, color="blue")
drawing.oval(30, 50, 290, 190, color='#ff2277')
drawing.line(0, 0, 400, 200, color='black', width=5)
drawing.text(20, 100, "Hello World", color="green",
    font="Times", size=48)
app.display()
```

The Drawing is, in this case, created with a width and height of fill so that the Drawing fills the window. Although the program writes some text on the drawing, this works quite differently than using a Text. Whereas we can change the value of a Text and see that change on the screen, any text that we put on the drawing cannot be modified, unless you want to try carefully writing over the same area of the drawing.

Pop-Ups

Often when you are designing a program, you want a pop-up notification to appear over the window. Or, you might need to ask the user a question such as

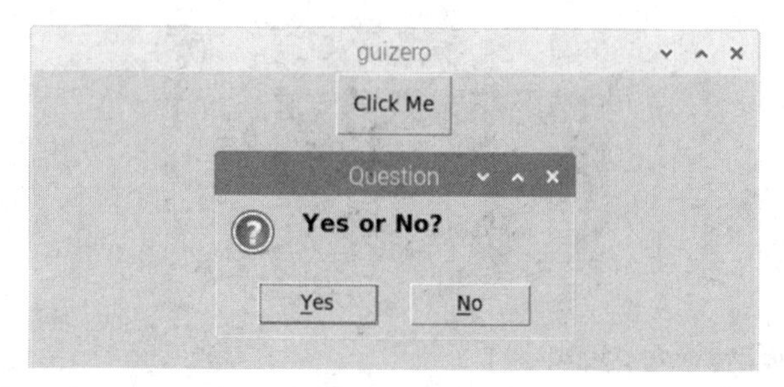

Figure 7-6 *The yes/no pop-up.*

confirming some action. There is a big selection of such pop-ups that you can use in guizero. Figure 7-6 shows the common yes/no type pop-up.

And here is the code for it:

```
#07_06_yes_no.py

from guizero import *

def ask():
    if yesno("Question", "Yes or No?"):
        info("Result", "You clicked Yes")
    else:
        warn("Result", "You clicked No")

app = App()
button = PushButton(app, text="Click Me", command=ask)
app.display()
```

The code actually demonstrates three types of pop-up. The main window contains only a button, which when pressed runs the function `ask`. This uses the first type of pop-up (`yesno`) to display the small window shown in Figure 7-6 that has a title of Question and a message "Yes or No?". The function `yesno` returns a Boolean value: true if Yes was clicked, otherwise false. If Yes is clicked then the pop-up info is used to display a second message "You clicked Yes". If, on the other hand, No is clicked then the warn pop-up is used to display the message "You clicked No". The difference between info and warn being that warn also displays a warning icon with the message.

There are a number of other types of pop-up. They are listed below and you can find full documentation on using them here: https://lawsie.github.io/guizero/alerts/.

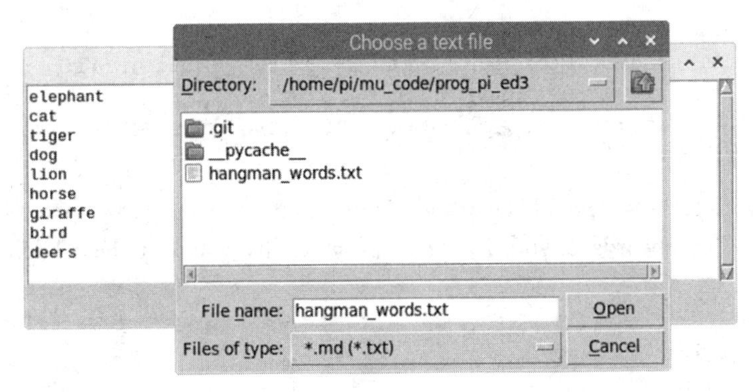

Figure 7-7 *The select_file dialog.*

- warn
- info
- error
- yesno—returns True or False
- question—returns the string typed by the user in response to a question
- select_file—allows the selection of a file
- select_folder – allows you to select a folder

Here's an example that uses the select_file dialog and then reads that file, displaying its contents using a TextBox. You can see this in action in Figure 7-7.

```
#07_07_file_viewer

from guizero import *

def ask():
    filename = select_file(title="Choose a text file",
        filetypes=[["*.md", "*.txt"]])
    if not filename:
        print("No file selected")
    else:
        read_file(filename)

def read_file(filename):
    f = open(filename)
    text = f.read()
    f.close()
    text_area.value = text
```

```
app = App(width=600, height=200)
text_area = TextBox(app, width="fill", height=10,
    multiline=True, scrollbar=True)
button = PushButton(app, text="Open", command=ask)
app.display()
```

As an exercise, you might like to add a Save button to this program so that you can use it to edit the words for your Hangman program that you started in Chapter 4.

Menus

The guizero also provides us with a way of adding menus to our application. Figure 7-8 shows a modified version of our file viewer program that has replaced the Open button with a menu. We've also added some extra options to the menu (Save and Quit).

If you want to try this app out for yourself, you will find it in 07_08_file_viewer_menu.py. The main are a of interest in the code where the menu bar for the window is defined with the app:

```
menubar = MenuBar(app,
    toplevel=["File", "Edit"],
    options=[
        [["Open", ask_file], ["Save", save_file],
            ["Quit", quit_app]],
        [["Find", find]]
            ])
```

The MenuBar control has two parameters. The first of these (`toplevel`) contains a list of strings that are the names of the menus to appear in the menu bar at

Figure 7-8 *Menus in guizero.*

the top of the window. The second parameter (`options`) contains the individual menu options grouped into a list for each menu name. Each of these options is itself a list containing the name of the option to be displayed and the name of the function to be called if that menu option is selected.

Notice how the Quit menu option calls the method `app.destroy` to close the window.

Summary

There is lots more to the guizero library, so I would encourage you to browse the documentation that you can find here: https://lawsie.github.io/guizero.

You will use guizero in various other places in this book.

8

Games Programming

Clearly a single chapter is not going to make you an expert in game programming. A number of good books are devoted specifically to game programming in Python, such as *Beginning Game Development with Python and Pygame*, by Will McGugan. This chapter introduces you to a very handy library called pygame and gets you started using it to build a simple game.

What Is Pygame?

Pygame is a library that makes it easier to write games for the Raspberry Pi—or more generally for any computer running Python. The reason why a library is useful is that most games have certain elements in common, and you'll encounter some of the same difficulties when writing them. A library such as pygame takes away some of this pain because someone really good at Python and game programming has created a nice little package to make it easier for us to write games. In particular, pygame helps us in the following ways:

- We can draw graphics that don't flicker.
- We can control the animation so that it runs at the same speed regardless of whether we run it on a Raspberry Pi or a top-of-the-range gaming PC.
- We can catch keyboard and mouse events to control the game play.

Coordinates

When using guizero, we used a grid layout to control where things appeared in the window, so you never needed to worry about the exact positions of things. In

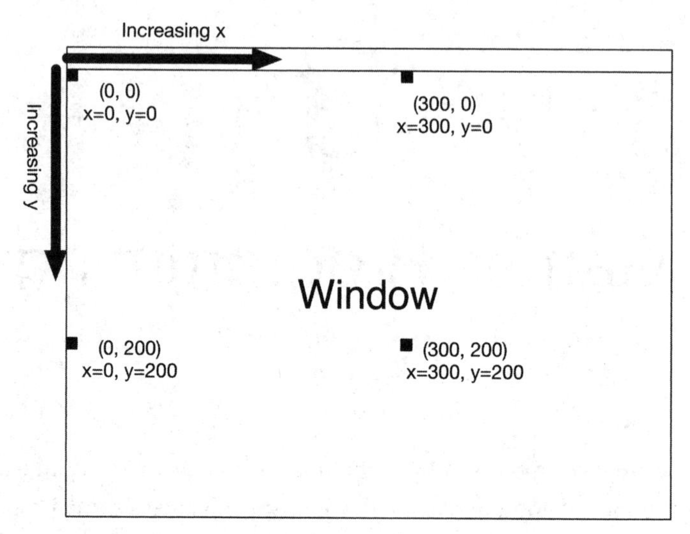

Figure 8-1 *The pygame coordinate system.*

pygame, coordinates are specified as values of X and Y relative to the top left corner of the window. X values are left to right and Y are from top to bottom. Figure 8-1 illustrates the pygame coordinate system.

Coordinates are often written as a tuple with the X value first. So (100, 200) refers to the point X=100, Y=200.

Hello Pygame

Figure 8-2 shows what a Hello World–type application looks like in pygame, and here is the code listing for it:

```
#08_01_hello_pygame.py

import pygame

pygame.init()

screen = pygame.display.set_mode((200, 200))
screen.fill((255, 255, 255))
pygame.display.set_caption('Hello Pygame')
```

```
ball = pygame.image.load('prog_pi_ed3/raspberry.jpg').
    convert()
screen.blit(ball, (100, 100))

pygame.display.update()
```

Figure 8-2 *Hello Pygame.*

This is a very crude example, and it doesn't have any way of exiting gracefully. Clicking on the Sop button in Mu should kill it after a few seconds.

Looking at the code for this example, you can see that the first thing we do is import `pygame`. The method `init` (short for *initialize)* is then run to get pygame set up and ready to use. We then assign a variable called `screen` using the line

```
screen = pygame.display.set_mode((200, 200))
```

which creates a new window that's 200 by 200 pixels. We then fill it with white (the color 255, 255, 255) on the next line before setting a caption for the window of "Hello Pygame."

Games use graphics, which usually means using images. In this example, we read an image file into pygame:

```
raspberry = pygame.image.load('prog_pi_ed3/raspberry.
    jpg').convert()
```

In this case, the image is a file called raspberry.jpg, which is included along with all the other programs in this book in the programs download section on the book's website. The call to `convert()` at the end of the line is important because it converts the image into an efficient internal representation that enables it to be drawn very quickly, which is vital when we start to make the image move around the window.

Next, we draw the raspberry image on the screen at coordinates 100, 100 using the `blit` command. As with the guizero canvas you met in the previous chapter, the coordinates start with 0, 0 in the top-left corner of the screen.

Finally, the last command tells pygame to update the display so that we get to see the image.

A Raspberry Game

To show how pygame can be used to make a simple game, we are going to gradually build up a game where we catch falling raspberries with a spoon. The raspberries fall at different speeds and must be caught on the eating end of the spoon before they hit the ground. Figure 8-3 shows the finished game in action. It's simple but functional. Hopefully, you will take this game and improve upon it.

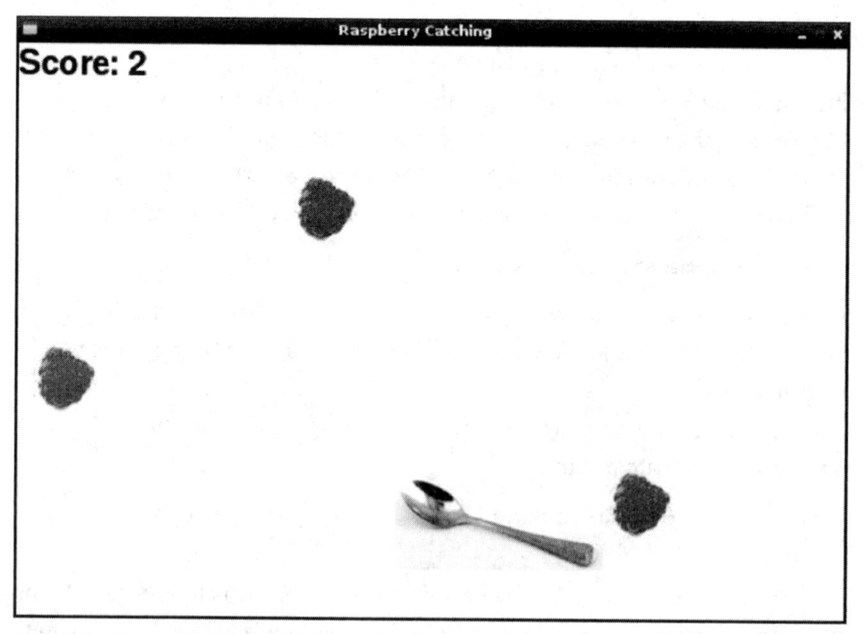

Figure 8-3 *The raspberry game.*

Following the Mouse

Let's start developing the game by creating the main screen with a spoon on it that tracks the movements of the mouse left to right. Load the following program into Mu:

```
#08_02_rasp_game_mouse

import pygame
from pygame.locals import *

spoon_x = 300
spoon_y = 300

pygame.init()

screen = pygame.display.set_mode((600, 400))
pygame.display.set_caption('Raspberry Catching')

spoon = pygame.image.load('spoon.jpg').convert()

while True:

    for event in pygame.event.get():
        if event.type == QUIT:
            pygame.quit()

    screen.fill((255, 255, 255))
    spoon_x, ignore = pygame.mouse.get_pos()
    screen.blit(spoon, (spoon_x, spoon_y))

    pygame.display.update()
```

The basic structure of our Hello World program is still there, but you have some new things to examine. First of all, there is another import. The import for `pygame.locals` provides us access to useful constants such as `QUIT`, which we will use to detect when the game is about to exit.

We have added two variables (`spoon_x` and `spoon_y`) to hold the position of the spoon. Because the spoon is only going to move left to right, `spoon_y` will never change.

At the end of the program is a `while` loop. Each time around the loop, we first check for a `QUIT` event coming from the pygame system. Events occur every time the player moves the mouse or presses or releases a key. In this case, we are only interested in a QUIT event, which is caused by someone clicking the window close icon in the top-right corner of the game window. We could choose not to exit immediately here, but rather prompt the player to see whether they indeed want to exit. The next line clears the screen by filling it with the color white.

Next comes an assignment in which we set spoon_x to the value of the x position of the mouse. Note that although this is a double assignment, we do not care about the y position of the mouse, so we ignore the second return value by assigning it to a variable called ignore that we then ignore. We then draw the spoon on the screen and update the display.

Run the program. The spoon should now follow the mouse movements.

One Raspberry

The next step in building the game is to add a raspberry. Later on we will expand this so that there are three raspberries falling at a time, but starting with one is easier. The code listing for this can be found in the file 08_03_rasp_game_one.py.

Here are the changes from the previous version:

- Add global variables for the position of the raspberry (raspberry_x and raspberry_y).
- Load and convert the image raspberry.jpg.
- Separate updating the spoon into its own function.
- Add a new function called update_raspberry.
- Update the main loop to use the new functions.

You should already be familiar with the first two items in this list, so let's start with the new functions:

```
def update_spoon():
    global spoon_x
    global spoon_y
    spoon_x, ignore = pygame.mouse.get_pos()
    screen.blit(spoon, (spoon_x, spoon_y))
```

The function update_spoon just takes the code we had in the main loop in 08_02_rasp_game_mouse and puts it in a function of its own. This helps to keep the size of the main loop down so that it is easier to tell what's going on.

```
def update_raspberry():
    global raspberry_x
    global raspberry_y
    raspberry_y += 5
    if raspberry_y > spoon_y:
        raspberry_y = 0
        raspberry_x = random.randint(10, screen_width)
    raspberry_x += random.randint(-5, 5)
    if raspberry_x < 10:
        raspberry_x = 10
    if raspberry_x > screen_width - 20:
        raspberry_x = screen_width - 20
    screen.blit(raspberry, (raspberry_x, raspberry_y))
```

The function `update_raspberry` changes the values of `raspberry_x` and `raspberry_y`. It adds 5 to the y position to move the raspberry down the screen and moves the x position by a random amount between –5 and +5. This makes the raspberries wobble unpredictably during their descent. However, the raspberries will eventually fall off the bottom of the screen, so once the y position is greater than the position of the spoon, the function moves them back up to the top and to a new random x position.

There is also a danger that the raspberries may disappear off the left or right side of the screen. Therefore, two further tests check that the raspberries aren't too near the edge of the screen, and if they are then they aren't allowed to go any further left or right.

Here's the new main loop that calls these new functions:

```
while True:
    for event in pygame.event.get():
        if event.type == QUIT:
            pygame.quit()

    screen.fill((255, 255, 255))
    update_raspberry()
    update_spoon()
    pygame.display.update()
```

Try out 08_03_rasp_game_one.py. You will see a basically functional program that looks like the game is being played. However, nothing happens when you catch a raspberry.

Catch Detection and Scoring

We are now going to add a message area to display the score (that is, the number of raspberries caught). To do this, we must be able to detect that we have caught a raspberry. The extended program that does this is in the file 08_04_rasp_py_game_scoring.py.

The main changes for this version are two new functions, `check_for_catch` and `display`:

```
def check_for_catch():
    global score
    if raspberry_y >= spoon_y and raspberry_x >= spoon_x and \
       raspberry_x < spoon_x + 50:
        score += 1
    display("Score: " + str(score))
```

Note that because the condition for the `if` is so long, we use the line-continuation command (\) to break it into two lines.

The function `check_for_catch` adds 1 to the score if the raspberry has fallen as far as the spoon (`raspberry_y >= spoon_y`) and the x position of the raspberry is between the x (left) position of the spoon and the x position of the spoon plus 50 (roughly the width of the business end of the spoon).

Regardless of whether the raspberry is caught, the score is displayed using the `display` function. The `check_for_catch` function is also added into the main loop as one more thing we must do each time around the loop.

The `display` function is responsible for displaying a message on the screen.

```
def display(message):
    font = pygame.font.Font(None, 36)
    text = font.render(message, 1, (10, 10, 10))
    screen.blit(text, (0, 0))
```

You write text on the screen in pygame by creating a font, in this case, of no specific font family but of a 36-point size and then create a `text` object by rendering the contents of the string `message` onto the font. The value `(10, 10, 10)` is the text color. The end result contained in the variable `text` can then be blitted onto the screen in the usual way.

Timing

You may have noticed that nothing in this program controls how fast the raspberries fall from the sky. You may have noticed that the raspberries fall quite quickly.

The speed they fall at will also depend on how fast our Raspberry Pi is, so a Raspberry Pi 4 will be much faster than a Raspberry Pi 1.

To manage the speed, pygame has a built-in clock that allows us to slow down our main loop by just the right amount to perform a certain number of refreshes per second. Unfortunately, it can't do anything to speed up our main loop. This clock is very easy to use; you simply put the following line somewhere before the main loop:

```
clock = pygame.time.Clock()
```

This creates an instance of the clock. To achieve the necessary slowing of the main loop, put the following line somewhere in it (usually at the end):

```
clock.tick(30)
```

In this case, we use a value of `30`, meaning a frame rate of 30 frames per second. You can put a different value in here, but the human eyes (and brain) do not register any improvement in quality above about 30 frames per second.

Lots of Raspberries

Our program is starting to look a little complex. If we were to add the facility for more than one raspberry at this stage, it would become even more difficult to see what is going on. We are therefore going to perform *refactoring,* which means changing a perfectly good program and altering its structure without changing what it actually does or without adding any features. We are going to do this by creating a class called `Raspberry` to do all the things we need a raspberry to do. This still works with just one raspberry, but will make working with more raspberries easier later. The code listing for this stage can be found in the file 08_05_rasp_game_refactored.py. Here's the class definition:

```
class Raspberry:
    x = 0
    y = 0

    def __init__(self):
        self.x = random.randint(10, screen_width)
        self.y = 0

    def update(self):
        self.y += 5
        if self.y > spoon_y:
            self.y = 0
            self.x = random.randint(10, screen_width)
        self.x += random.randint(-5, 5)
```

```
        if self.x < 10:
            self.x = 10
        if self.x > screen_width - 20:
            self.x = screen_width - 20
        screen.blit(raspberry_image, (self.x, self.y))

    def is_caught(self):
        return self.y >= spoon_y and self.x >= spoon_x and \
                self.x < spoon_x + 50
```

The raspberry_x and raspberry_y variables just become variables of the new Raspberry class. Also, when an instance of a raspberry is created, its x position will be set randomly. The old update_raspberry function has now become a method on Raspberry called just update. Similarly, the check_for_catch function now asks the raspberry if it has been caught.

Having defined a raspberry class, we create an instance of it like this:

```
r = Raspberry()
```

Thus, when we want to check for a catch, the check_for_catch just asks the raspberry like this:

```
def check_for_catch():
    global score
    if r.is_caught():
        score += 1
```

The call to display the score has also been moved out of the check_for_catch function and into the main loop. With everything now working just as it did before, it is time to add more raspberries. The final version of the game can be found in the file 08_06_rasp_game_final.py. It is listed here in full:

```
#08_06_rasp_game_final

import pygame
from pygame.locals import *
import random

score = 0

screen_width = 600
screen_height = 400

spoon_x = 300
spoon_y = screen_height - 100

class Raspberry:
    x = 0
```

```
    y = 0
    dy = 0

    def __init__(self):
        self.x = random.randint(10, screen_width)
        self.y = 0
        self.dy = random.randint(3, 10)

    def update(self):
        self.y += self.dy
        if self.y > spoon_y:
            self.y = 0
            self.x = random.randint(10, screen_width)
        self.x += random.randint(-5, 5)
        if self.x < 10:
            self.x = 10
        if self.x > screen_width - 20:
            self.x = screen_width - 20
        screen.blit(raspberry_image, (self.x, self.y))

    def is_caught(self):
        return self.y >= spoon_y and self.x >= spoon_x

              and self.x < spoon_x + 50

clock = pygame.time.Clock()
rasps = [Raspberry(), Raspberry(), Raspberry()]

pygame.init()

screen = pygame.display.set_mode((screen_width, screen_height))
pygame.display.set_caption('Raspberry Catching')

spoon = pygame.image.load('prog_pi_ed3/spoon.jpg').convert()
raspberry_image = pygame.image.load('prog_pi_ed3/raspberry.jpg').
convert()

def update_spoon():
    global spoon_x
    global spoon_y
    spoon_x, ignore = pygame.mouse.get_pos()
    screen.blit(spoon, (spoon_x, spoon_y))

def check_for_catch():
    global score
    for r in rasps:
        if r.is_caught():
            score += 1

def display(message):
    font = pygame.font.Font(None, 36)
    text = font.render(message, 1, (10, 10, 10))
    screen.blit(text, (0, 0))
```

```
while True:
    for event in pygame.event.get():
        if event.type == QUIT:
            pygame.quit()

    screen.fill((255, 255, 255))
    for r in rasps:
        r.update()
    update_spoon()
    check_for_catch()
    display("Score: " + str(score))
    pygame.display.update()
    clock.tick(30)
```

To create multiple raspberries, the single variable `r` has been replaced by a collection called `rasps`:

```
rasps = [Raspberry(), Raspberry(), Raspberry()]
```

This creates three raspberries; we could change it dynamically while the program is running by adding new raspberries to the list (or for that matter removing some).

We now need to make just a couple other changes to deal with more than one raspberry. First of all, in the `check_for_catch` function, we now need to loop over all the raspberries and ask each one whether it has been caught (rather than just the single raspberry). Second, in the main loop, we need to display all the raspberries by looping through them and asking each to update.

Summary

You can learn plenty more about pygame. The official website at www.pygame.org has many resources and sample games that you can play with or modify.

9

Interfacing Hardware

The Raspberry Pi has a double row of pins on one side of it. These pins are called the GPIO (General Purpose Input/Output) connector and allow you to connect electronic hardware to the Pi as an alternative to using the USB port.

The maker and education communities have created many expansion and prototyping boards you can attach to your Pi so you can add your own electronics. This includes everything from simple temperature sensors to relays. You can even convert your Raspberry Pi into a controller for a robot.

In this chapter, we explore the various ways of connecting the Pi to electronic devices using the GPIO connector. Because this is a fast-moving field, it is fairly certain that new products will have come on the market since this chapter was written; therefore, check the Internet to see what is current. I have tried to choose a representative set of different approaches to interfacing hardware. Therefore, even if the exact same versions are not available, you will at least get a flavor of what is out there and how to use it.

GPIO Pin Connections

All versions of Raspberry Pi since the Raspberry Pi 2 have two rows of 20 pins, making 40 pins in all, whereas the original Raspberry Pi has just 26 pins on the GPIO header. To maintain compatibility, the first 26 pins of the Raspberry Pi 2 and later are the same as the pins of the older Raspberry Pi. In other words, the Raspberry Pi 2, 3, and 4 give you some extra pins to use.

In Figure 9-1, the GPIO pins of the Raspberry Pi 4 have a GPIO template over the pins that labels each of the pins. The template shown in Figure 9-1 is the Raspberry Leaf. Other templates are also available.

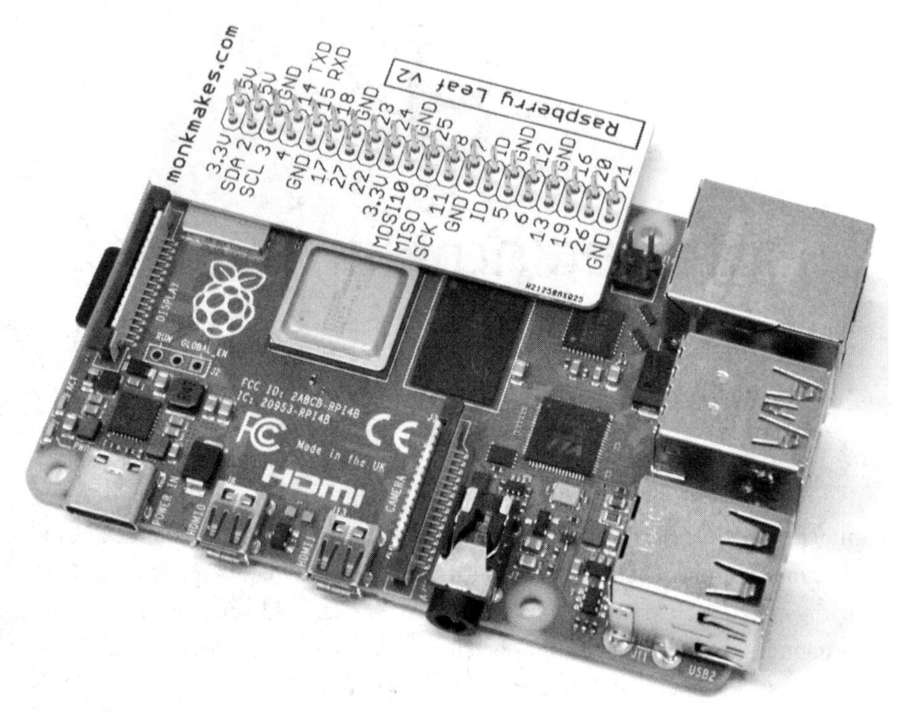

Figure 9-1 *Raspberry Pi model 4.*

Pin Functions

Figure 9-2a shows the pin names for the Raspberry Pi 2 and later, which are also the same as for the Raspberry Pi B+ and A+ and will probably remain much the same for any new models of Raspberry Pi that are released. Figure 9-2b shows the pin names for the older Raspberry Pi models A and B.

The pins labeled with a number can all be used as general-purpose input/output pins. In other words, any one of them can first be set to either an input or an output. If the pin is set to be an input, you can then test to see whether the pin is set to a "1" (above about 1.7V) or a "0" (below 1.7V). Note that all the GPIO pins are 3.3V pins and connecting them to higher voltages than that could damage your Raspberry Pi.

When set to be an output, the pin can be either 0V or 3.3V (logical 0 or 1). Pins can only supply or sink a small amount of current (assume 3mA to be safe), so they can just light an LED if you use a high-value resistor (say, 470Ω or higher).

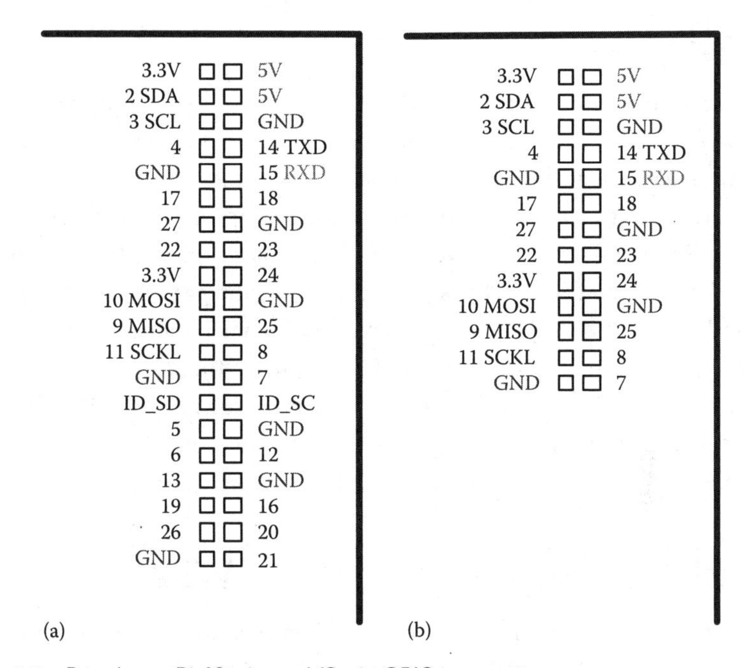

Figure 9-2 *Raspberry Pi 40-pin and 13-pin GPIO connectors.*

Serial Interface Pins

You will notice that some of the GPIO pins have other letters after their names. Those pins can be used as normal GPIO pins, but also have some special purpose. For example, pins 2 and 3 have the extra names of SDA and SCL. These are the clock and data lines, respectively, for a serial bus type called I2C that is popular for communicating with peripherals such as temperature sensors, LCD displays, and the like.

GPIO pins 14 and 15 also double as the TXD and RXD (Transmit and Receive) pins for the Raspberry Pi's serial port. Yet another type of serial communication is possible through GPIO 9 to 11 (MISO, MOSI, and SCLK). This type of serial interface is called SPI.

Power Pins

Both GPIO connectors are sprinkled with pins labeled GND (ground). These pins are all connected to the Raspberry Pi's ground or zero volts. Other power pins are also provided for 3.3V and 5V. You will often use these pins when hooking up external electronics to the Raspberry Pi.

Hat Pins

Two special pins, only available on the 40-pin variant of the Raspberry Pi, are ID_SD and ID_SC. These are reserved for an advanced interface standard that you can use with the Raspberry Pi 2 and later, B+ and A+. The standard is called HAT (Hardware Attached to Top). This standard does not in any way stop you just using GPIO pins directly; however, interface boards that conform to the HAT standard can call themselves HATs and have the advantage that a HAT must contain a little EEPROM (Electrically Erasable Programmable Read-Only Memory) chip on it that is used to identify the HAT so that ultimately the Raspberry Pi could auto-install necessary software. At the time of writing, HATs have not quite met that level of sophistication, but the idea is a good one. The pins ID_SD and ID_SC are used to communicate with a HAT EEPROM.

Breadboarding with Jumper Wires

Solderless breadboard, often just called breadboard, is a great way of connecting electronics to a Raspberry Pi. There is no soldering to do—you just push electronic components into the breadboard and then connect them to your Raspberry Pi GPIO connector using special jumper wires from the Raspberry Pi onto the breadboard.

Digital Outputs

A nice starting point with the GPIO connector is to wire-up an LED so that it can be turned on and off from a Python program. To wire-up the LED you will need the following items.

Part	Suppliers
Solderless breadboard	Adafruit (Product 64), SparkFun (SKU PRT-00112)
Female-to-male jumper wires	Adafruit (1954)
Red LED	Adafruit (299)
470Ω resistor (A 1kΩ resistor will also work)	MCM Electronics (34-470)

It's often easier to buy an electronics starter kit that has the common parts listed above. The MonkMakes Project Box 1 for Raspberry Pi contains all the parts listed above and a Raspberry Leaf for easy pin identification. You will also find starter kits on eBay that contain a wide range of components, to get you started.

Warning: Keeping Your Pi Safe

The Raspberry Pi has relatively delicate GPIO pins. It is possible to burn out individual GPIO pins and even destroy the whole Raspberry Pi if you are not careful.

Always check over the wiring carefully before connecting your electronics to your Raspberry Pi. In particular, make sure that you always use a resistor of at least 470Ω between a GPIO pin and an LED. The resistor limits the current through the LED to a safe level for the Raspberry Pi.

Step 1. Put the Resistor on the Breadboard

Breadboard is arranged in rows and columns. The rows are numbered 1 to 30 and the columns "a" to "j" (in two banks). All the holes for a particular row in a bank ("a" to "e" or "f" to "j") are connected together behind the plastic front of the breadboard by a metal clip. So putting two component legs into the same row connects them together electrically.

Start by putting the legs of the resistor both on column "c" between rows 1 and 6 as shown in Figure 9-3. It does not matter which way around the resistor goes.

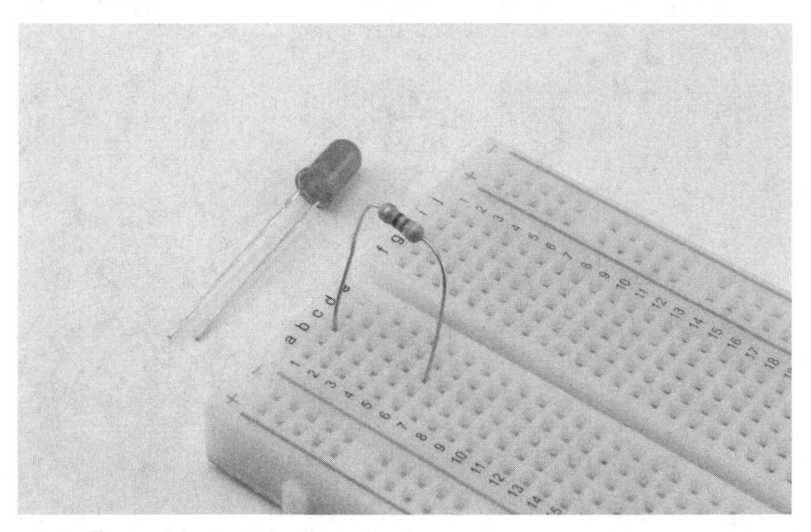

Figure 9-3 *The resistor on breadboard.*

Step 2. Put the LED on the Breadboard

The LED has one leg longer than the other. The longer leg is the positive leg and this should go to row 6, column "e" to connect to the bottom lead of the resistor. The other leg of the LED (the shorter lead) plugs into row 8, column "e" as shown in Figure 9-4.

Step 3. Connect the Breadboard to the GPIO Pins

You will need two female-to-male jumper wires. The male end will plug into the breadboard and the female end onto a GPIO pin. Pick different colors. I used black and orange. Plug one lead (let's say its orange) from row 1, column "a" to pin GPIO18 of the GPIO header. This pin is the sixth pin down on the right (see Figure 9-2). If you have a GPIO template like the Raspberry Leaf, it's a lot easier to see where to make the connection.

The other jumper wire needs to go from row 8, column "a" of the breadboard to one of the GND connections on the GPIO connector of the Pi. I used the GND pin that is the third pin down on the right hand side of the GPIO connector as shown in Figure 9-5.

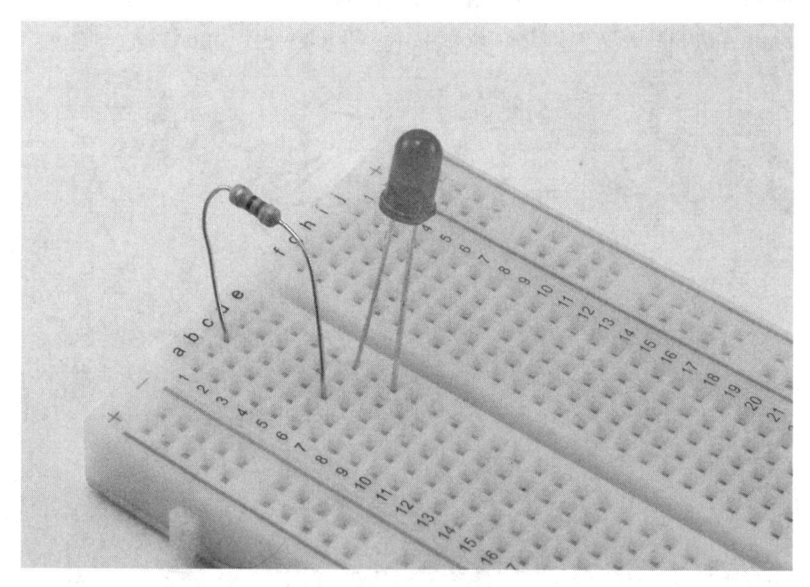

Figure 9-4 *The resistor and LED on breadboard.*

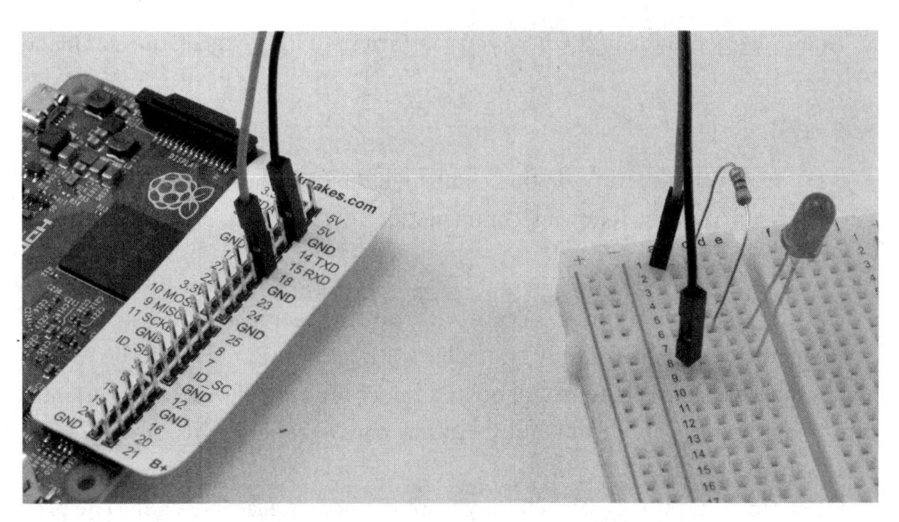

Figure 9-5 *The breadboard and Pi connected.*

Now that the LED is connected, you can try out some Python to turn it on and off. While you are trying this out, you will just enter commands in the Mu REPL or the Python 3 console. Here, I'll assume that you are opening the Python 3 console from the Terminal.

```
$ python3
Python 3.7.3 (default, Dec 20 2019, 18:57:59)
[GCC 8.3.0] on linux
Type "help", "copyright", "credits" or "license" for
more information.
>>>
```

To access the GPIO pins, you need to import a library called gpiozero. This library is included with Raspberry Pi OS so you do not need to install it. You just need to import it into the console by entering the command below:

```
import gpiozero
>>>
```

The LED is connected to pin GPIO 18, but at the moment the guizero library does not know if this pin should be an input or output. The following line specifies that it should be an output connected to an LED and give it the name `'led'`:

```
>>> led = gpiozero.LED(18)
```

At last, we get to the part where you can turn the LED on using the command below:

```
>>> led.on()
```

As soon as you hit return on that command, the LED should light. To turn the LED off again, type the following command:

```
>>> led.off()
```

Try this out a few times, because it's fun. It's actually quite significant, because although you are only controlling a humble LED, it could be a relay switching a domestic light on and off and the Python could be a home automation program. An important link between hardware and software has been established.

Open the program 09_01_blink.py from the book's example code. The program should be in the directory /home/pi/mu_code/prog_pi_ed3 if you followed the instructions for installing all the example programs back in Chapter 3. Run the program using Mu and you will see the LED start to blink on and off. When you have had enough, select the IDLE console window and press CTRL-C.

Here is the listing for 09_blink.py.

```
#09_01_blink.py

import gpiozero, time

led = gpiozero.LED(18)

while True:
    led.on()
    time.sleep(0.5)                 # delay 0.5 seconds
    led.off()
    time.sleep(0.5)
```

The program starts the same way as our earlier experiments, by importing the library, and also the "time" library. A variable (led) is used for the pin to be used to drive the LED. This is then initialized to be an output.

The main part of the program is a loop that continues until the program exits. This first turns on led_pin, delays for half a second, and then turns it off again and waits another half second before repeating the whole loop again.

Note that here we have done the LED blinking the "hard way." The LED class has a method called "blink" that takes two parameters that specify how long the LED should be on and how long off in each blinking cycle.

So, we could make the LED blink like this:

```
#09_02_blink_easy.py

import gpiozero, time

led = gpiozero.LED(18)

led.blink(on_time=0.5, off_time=0.5)
```

In this case, the program will actually continue after setting the LED blinking, which doesn't matter if you are running it from Mu, but if running it from the terminal, add the line `'input()'` to the end of the program to keep the program running until we press Enter. Otherwise the program will exit immediately and the blinking cease before it ever gets a chance to start.

Analog Outputs

Don't dismantle the LED and resistor breadboard just yet, because as well as turning the LED on and off, you can also vary its brightness.

Pulse Width Modulation

The method used by the gpiozero library to produce an "analog" output is called Pulse Width Modulation (PWM). The GPIO pin actually uses a digital output, but generates a series of pulses. The width of the pulses are varied. The larger the proportion of the time that the pulse stays high, the greater the power delivered to the output, and hence the brighter the LED as shown in Figure 9-6.

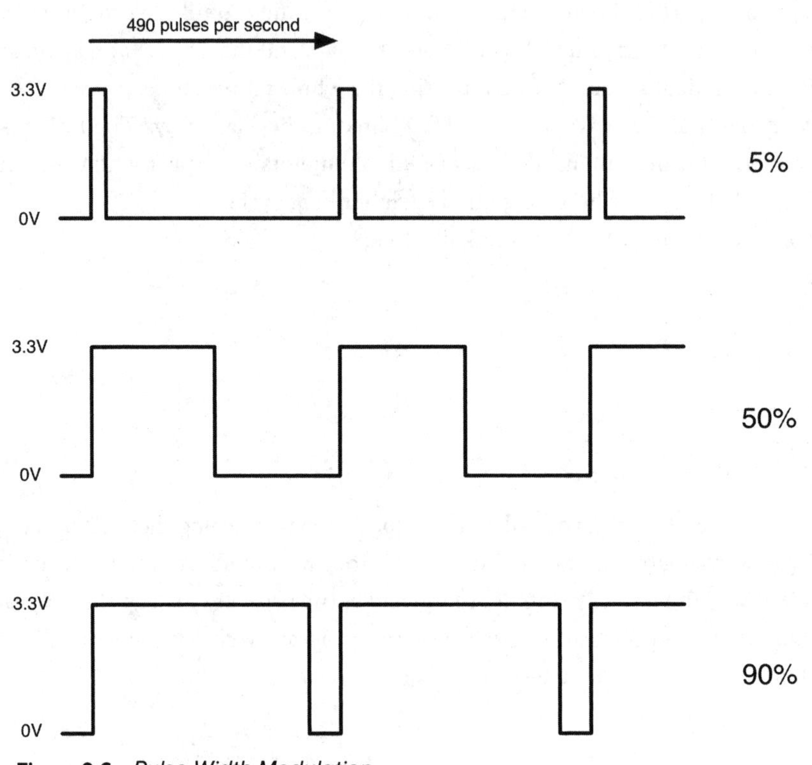

Figure 9-6 *Pulse Width Modulation.*

The proportion of the time that the pulse is HIGH is called the duty and this is often expressed as a percentage.

Even though the LED is actually turning on and off, it happens so fast that your eye is just fooled into thinking the LED is brighter or dimmer depending on the length of the PWM pulse.

With the LED connected to pin 18 as you did in Figure 9-5, open the program 09_03_pwm.py and then run it. The first thing the program does is to prompt you for a brightness level between 1 and 100 as shown below:

```
Enter Brightness (0 to 100):0
Enter Brightness (0 to 100):50
Enter Brightness (0 to 100):100
```

Try a few different values and see how the brightness of the LED changes.

Here's the code for this program:

```
#09_03_pwm.py

import gpiozero

led = gpiozero.PWMLED(18)

while True:
    duty_s = input("Enter Brightness (0 to 100):")
    duty = float(duty_s) / 100.0
    led.value = duty
```

Instead of using the class LED to control the LED, to be able to alter its brightness, you must use PWMLED. To change the brightness, you just set its value to a number between 0 and 1, where 0 is completely off and 1 is maximum brightness.

The main loop prompts you to enter the brightness as a string and then converts it to a number between 0 and 1 before setting the new level of brightness for the LED.

Digital Inputs

Where an LED is the most likely thing to be connected to a digital output, a switch is probably the most likely thing to be connected to a digital input.

To experiment with a switch as a digital input, you don't actually need a switch, or breadboard, you can just experiment with a pair of female-to-male jumper wires. Connect one wire to GND and the other to pin 23 as shown in Figure 9-7.

Open the program 09_04_switch.py in Mu and then run it. When you touch the wires together, a new line of output should appear in the console as shown in Figure 9-8.

Here is the listing for this program.

```
#09_04_switch.py

import gpiozero, time

switch = gpiozero.Button(23, pull_up=True)
```

```
while True:
    if switch.is_pressed:
        print("Button Pressed")
        time.sleep(0.2)
```

As with the previous blink program, there are the usual imports. A different pin is used for the switch. You could use pin 18 if you prefer, but by using 23, it leaves open the possibility of keeping the LED connected to pin 18 and combining both inputs and outputs.

This time, the gpiozero class we use is Button. Its first parameter is the pin to use and the parameter pull_up is set to True. This enables an internal pull-up resistor on pin 23 that keeps the input pulled up high unless it is connected to GND which overrides this.

The "while" loop contains an "if" statement that reads the digital input pin 23 using switch.is_pressed. If this is True, then it means that the input is connected to GND (the wires are connected) and the message is displayed.

The time.sleep command ensures that the messages don't go shooting off the screen when the wires are pressed, by introducing a one-fifth of a second delay before anything else happens in the loop.

Analog Inputs

Even the Raspberry Pi does not have analog inputs, that is inputs that can measure a voltage rather than simply tell if it is above or below a threshold that indicates the input is high or low. A lot of analog sensors provide an output voltage that is proportional to the thing they are measuring. So, for example, a temperature sensor chip such as the TMP36 has an output pin whose voltage varies depending on the temperature. The only way to use such a sensor with the Raspberry Pi is to use an ADC (Analog to Digital Convertor) chip.

However, many sensors are resistive. That is, their resistance changes with the thing they are measuring. A thermistor's resistance changes with temperature and a photoresistor's resistance varies depending on the amount of light falling on it. Other types of resistive sensors include gas sensors, strain sensors, and even resistive touch screens. These "resistive" sensors can be used with a Raspberry Pi by timing how long it takes for current to flow through the resistive sensor and charge up a capacitor to the extent that it crosses the threshold of a digital input so that the input counts as HIGH rather than LOW.

Figure 9-7 *Jumper wires as a switch.*

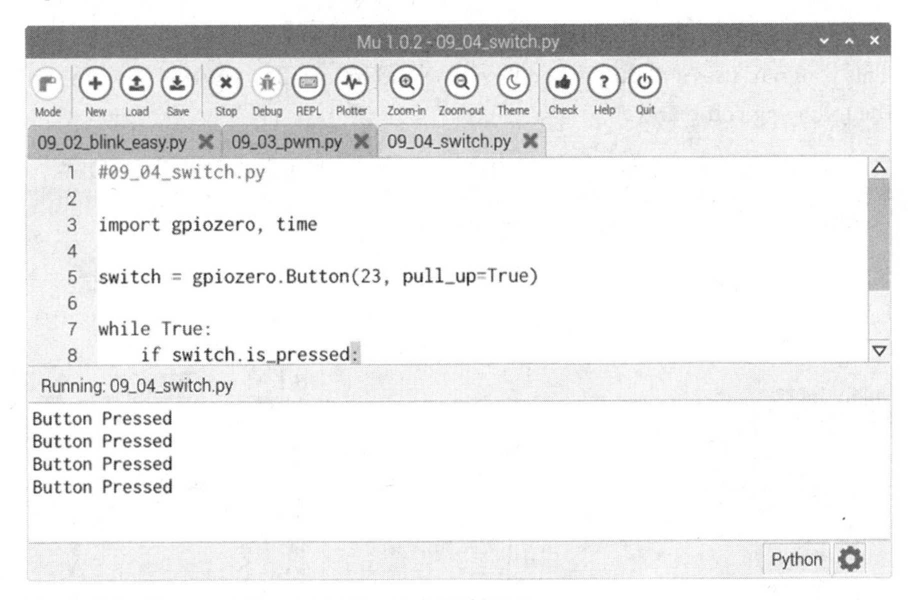

Figure 9-8 *The console monitoring a digital input.*

Hardware

You can try out this approach on breadboard using a photoresistor. To do this, you will need the following items:

- A half-sized breadboard (Adafruit PID: 64)

- Female-to-male jumper wires (Adafruit PID: 1954)
- Photoresistor (1kΩ) (Adafruit PID: 161)
- Two 1kΩ resistors (MCM Electronics PID: 66-1K)
- 330nF capacitor (MCM Electronics PID: 31-11864)

All these parts are also included in the MonkMakes Project Box 1 kit for Raspberry Pi.

Although a phototransistor is substituted for the photoresistor in this kit.

Figure 9-9 shows the wiring diagram for the breadboard. A photograph of the breadboard is shown in Figure 9-10.

None of the components need to be a particular way around. It can help to keep things neat if you shorten the length of the resistor legs before you fit them onto the breadboard.

The Software

This program uses a library called PiAnalog and to install it, you will need to run the following commands:

```
$ git clone https://github.com/simonmonk/pi_analog.git
$ cd pi_analog
$ sudo python3 setup.py install
```

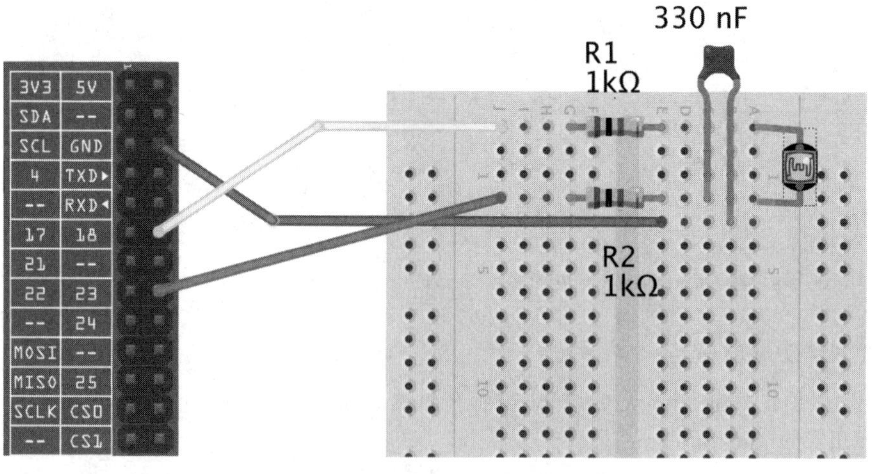

Figure 9-9 *The breadboard layout for light measurement.*

want to pick up *Python: Visual QuickStart Guide,* by Toby Donaldson. It's similar to this book in style, but provides a different perspective. Also, it's written in a friendly, reassuring manner. If you want something a bit more meaty, but still essentially a beginner's text, consider *Python Programming: An Introduction to Computer Science,* by John Zelle.

When it comes to learning more about pygame, you'll find *Beginning Game Development with Python and Pygame,* by Will McGugan, to be quite helpful.

Finally, here are some good web resources for Python you'll probably want to add to your browser's favorites list:

- **http://docs.python.org/py3k/** The official Python site, complete with useful tutorials and reference material.
- **https://lawsie.github.io/guizero** The official documentation for guizero.
- **https://gpiozero.readthedocs.io** The documentation for gpiozero.
- **www.pygame.org** The official pygame site. It contains news, tutorials, reference material, and sample code.

Raspberry Pi Resources

The official website of the Raspberry Pi Foundation is www.raspberrypi.org. This website contains a wealth of useful information, and it's the place to find announcements relating to happenings in the world of Raspberry Pi.

The forums are particularly useful when you are looking for the answer to some knotty problem. You can search the forum for information from others who have already tried to do what you are trying to do, you can post questions, or you can just show off what you've done to the community. When you're looking to update your Raspberry Pi distribution image, this is probably the best place to turn. The downloads page lists the distributions currently in vogue.

The Raspberry Pi even has its own online magazine, wittily named *The MagPi.* This is a free PDF download (www.themagpi.com) and contains a good mixture of features and "how-to" articles that will inspire you to do great things with your Pi.

For more information about the hardware side of using the Raspberry Pi, the following links are useful:

- **http://elinux.org/RPi_VerifiedPeripherals** A list of peripherals verified as working with the Raspberry Pi.

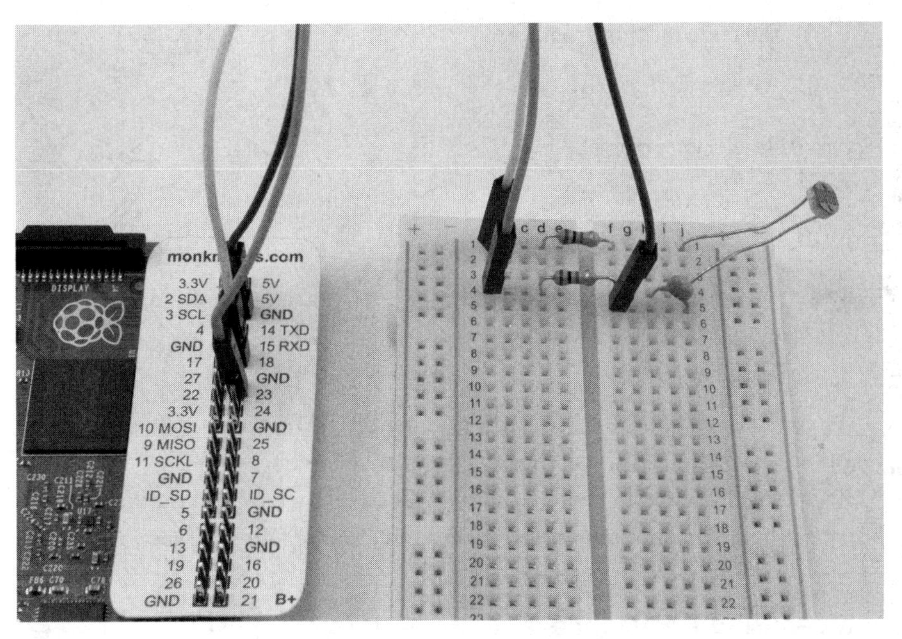

Figure 9-10 *Measuring light intensity with a photoresistor.*

The example code for reading resistance this way can be found in the file 09_05_resistance.py. When you run this program you will see output something like this in the console:

```
3648.04267883
3663.63811493
3608.03699493
10764.3079758
11204.1444778
11019.4854736
3608.94107819
3647.43995667
```

The increase in the resistance readings from around 3600 to 11000 occurred when I covered the photoresistor with my hand to make it darker.

You could swap the photoresistor for any other type of resistor or sensor to measure its value. Although this method is not very accurate, it can still be pretty useful.

Here's the code for this program:

```
#09_05_resistance.py

from PiAnalog import *
import time

p = PiAnalog()

while True:
    print(p.read_resistance())
    time.sleep(1)
```

As you can tell, all the clever stuff is going on inside the library. To read the resistance, we just need to call the read_resistance method.

If you are interested in how this library works, take a look at the github page at https://github.com/simonmonk/pi_analog and also the following section.

Measuring Resistance

To understand how this works, it can help to think of the capacitor as a water tank, the wires as pipes and the resistors, and the photoresistor as faucets that restrict the flow of water in the pipes.

First the capacitor is emptied of charge (the tank is emptied of water) by setting pin 23 to be an output and low. The charge then drains out of the capacitor through R2. R2 is there to make sure the charge doesn't flow out and into the Pi so fast that it damages the GPIO pin.

Next, pin 23 is effectively disconnected by setting it to be an input and pin 18 is set high (3.3V) so that the capacitor starts to fill through both the fixed resistor R1 and the photoresistor. The voltage at the capacitor will then start to rise as the capacitor fills. The capacitor will fill faster the lower the resistance of the photoresistor. This voltage is now monitored by pin 23 now acting as an input until the input goes high at about 1.65V (half of 3.3V). The time taken for this to happen is measured and can then be used to calculate the resistance of the photoresistor, which is an indication of the light level.

Figure 9-11a shows an oscilloscope trace of the voltage at pin 23 as the capacitor charges. The horizontal axis is time and the vertical axis volts. Figure 9-11b shows the same thing but with the photoresistor covered so that it is darker (and higher resistance). As you can see, it takes perhaps three times as long for the voltage to rise in the dark.

And now we get to the tricky math part. When a capacitor is charged through a resistor, the time taken for the capacitor voltage to rise to 0.632

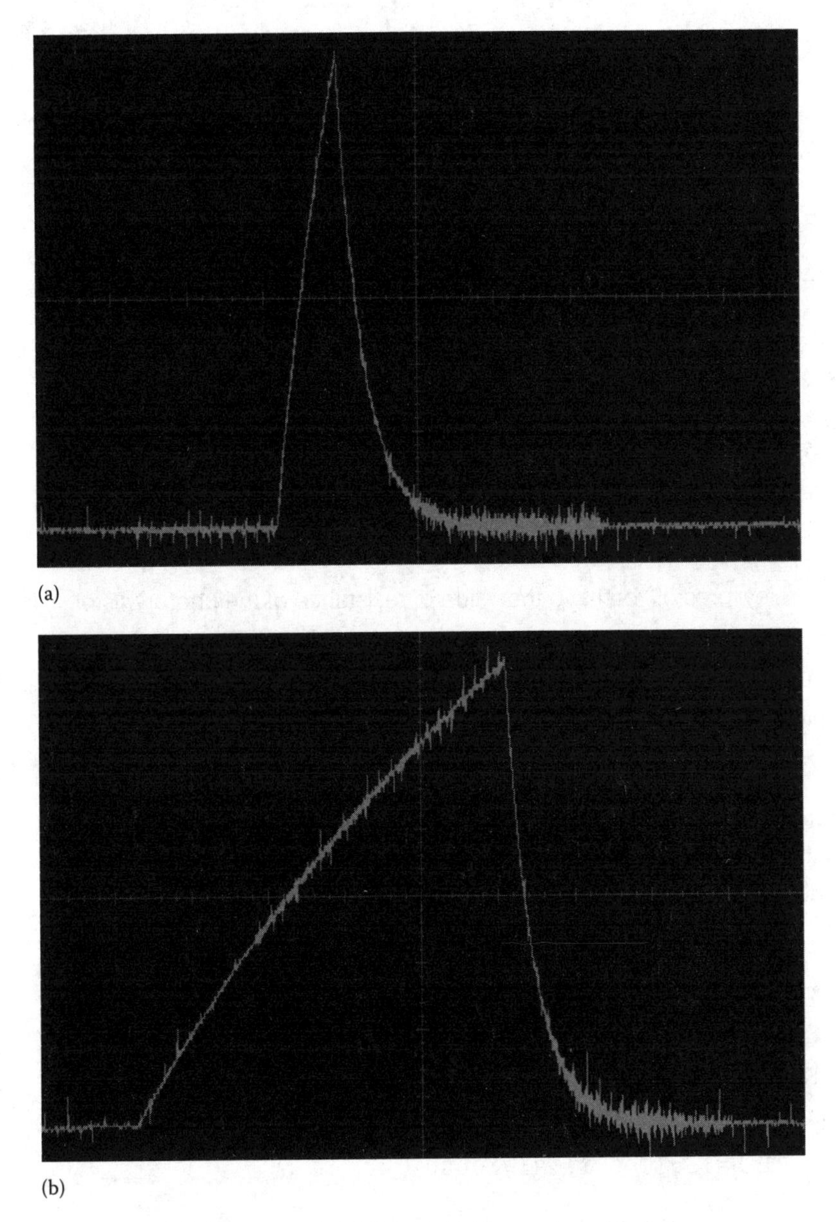

Figure 9-11 *(a) Voltage at pin 23 light. (b) Voltage at pin 23 dark.*

of the charging voltage is called the time constant (*T*). By the miracle of physics, *T* is also equal to the value of resistance times the capacitance.

So, you can work out *T* from the time taken to charge to 1.65V (*t*) using the equation:

$$T = t \times 3.3 \times 0.632$$

This is because we know how long it got to 1.65V we just need to scale that up a bit to see how long it would take to get to 3.33.3 × 0.632 = 2.09V. You now have a definite value for *T*.

Now you also know that:

$$T = (R + R1) \times C$$

where *R* is the photoresistor's resistance.

Rearranging these, you get:

$$R = (T/C) - R1$$

Hey, presto! You have the value of resistance of the photoresistor.

The PiAnalog library makes some assumptions about the values of resistor and capacitor that you are using. If you just use this in your code:

```
p = PiAnalog()
```

then it is assumed that you are using a 330nF capacitor and 1kΩ resistors. You can change these values by supplying optional parameters in the constructor. Those parameters are as follows:

- C—the capacitor value in μF
- R1—the value of R1 and R2 in Ωs
- Vt—the on voltage threshold

For example, if you were to use a 100nF capacitor and 10kΩ resistors you would use:

```
p = PiAnalog(C=0.1, R1=10000)
```

HATs

Another approach to using the Raspberry Pi's GPIO pins is to use a HAT (Hardware Attached Top). These are add-on boards that fit over the Raspberry Pi's GPIO pins.

Figure 9-12 *The Sense HAT.*

There are many interesting HATs available for the Raspberry Pi and the list is increasing in length all of the time. Some interesting board manufacturers to look for are Adafruit and Pimoroni who make and sell a wide variety of boards including displays, motor controllers, and touch sensing. You will also meet a motor controller board in Chapter 12.

Figure 9-12 shows a popular and useful Raspberry Pi Sense HAT. This has a display and a range of useful sensors as well as an easy-to-use Python library. You can find out more about the Sense HAT here: https://www.raspberrypi.org/products/sense-hat/.

Summary

In this chapter we looked at just some of the wide range of ways of adding electronics to our Raspberry Pi projects. In the next three chapters, we create projects using breadboard and jumper wires, and use a motor controller HAT as the basis for a small roving robot.

10

LED Fader Project

This is the first of three projects designed to make use of Python and the GPIO pins to control the color of light coming from an RGB LED. The project combines the use of the guizero library to create a user interface and the gpiozero libraries' PWM feature to control the brightness of the three channels of the LED (red, green, and blue).

Figure 10-1 shows the LED hardware built onto breadboard and Figure 10-2 shows the user interface used to control it on your Raspberry Pi.

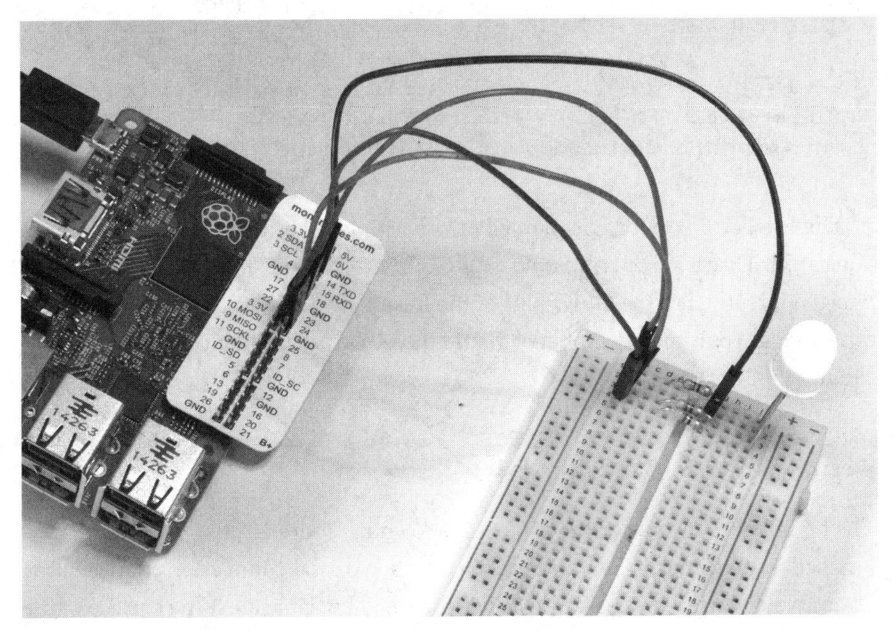

Figure 10-1 *An RGB LED connected to a Raspberry Pi.*

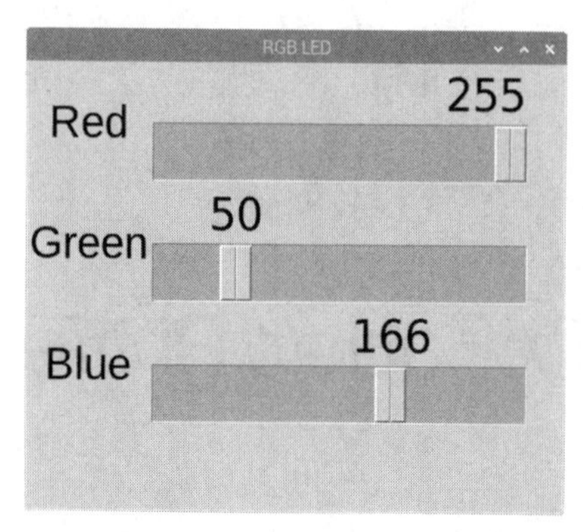

Figure 10-2 *A guizero user interface for controlling the LED.*

What You Need

To build this project, you will need the following parts. Suggested part suppliers are listed, but you can also find these parts elsewhere on the Internet.

Part	Suppliers
Solderless breadboard	Adafruit (Product 64), Sparkfun (SKU PRT-00112), Maplin (AG09K)
Female-to-male jumper wires	Adafruit (1954), Sparkfun (PRT-09385)
RGB common cathode LED	Sparkfun (COM-105)
3 × 470Ω resistor (1kΩ resistor will also work)	MCM Electronics (34-470)

The Project Box 1 for Raspberry Pi from MonkMakes includes all these parts. You can also use a Raspberry Squid, an RGB LED with built-in resistors that can be plugged directly into the GPIO pins of the Raspberry Pi. You can find instructions on making your own Raspberry Squid here: https://github.com/simonmonk/squid.

Hardware Assembly

The breadboard layout for the project is shown in Figure 10-3.

It will keep things neater and prevent any accidental connections between the leads if you shorten the resistor leads so that they lie flat against the surface of the breadboard.

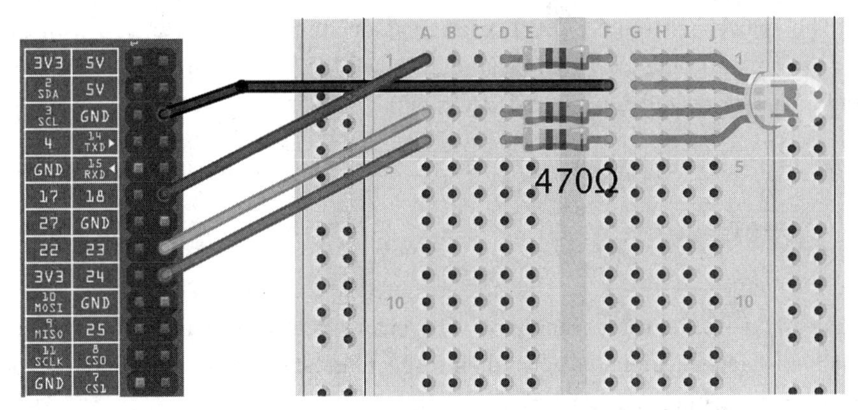

Figure 10-3 *The breadboard layout for an RGB LED.*

The RGB LED will have one leg that is longer than the others. This is the "common" lead. When you buy your RGB LED, make sure that it is specified as being "common cathode." This means that the negative terminals of each of the red, green, and blue LED elements are all connected together.

Software

The software for this project has some similarity with the experiment in Chapter 9, where you controlled the brightness of a single red LED by typing in a value between 0 and 100. However, in this project, instead of entering a number, guizero is used to create a user interface that has three sliders in a window. Each slider controls the brightness of a different channel, allowing you to mix red, green, and blue light to make any color.

Run the program and after a few moments the window shown in Figure 10-2 will appear. Try adjusting the sliders and notice how the LED color changes. LEDs with a diffuse body mix the colors much better than those with a clear body.

You can find the program in the book examples as the file 10_01_RGB_LED.py. Rather than list the whole program here, open it up in Mu while I go through the code in sections.

The program starts with the usual imports.

```
from gpiozero import RGBLED
from guizero import App, Slider, Text
from colorzero import Color
```

The gpiozero class RGBLED is used to associate pins 18, 23, and 24 with the red, green, and blue LED channels, respectively, and three variables red, green, and blue defined and given initial values of 0.

```
rgb_led = RGBLED(18, 23, 24)
red = 0
green = 0
blue = 0
```

Next, we have three functions, red_changed, green_changed, and blue_changed. Each will be called when their respective slider is moved. Here's the code for the red channel:

```
def red_changed(value):
    global red
    red = int(value)
    rgb_led.color = Color(red, green, blue)
```

This first sets the value of the global variable red to be the parameter passed to it after first converting it to a number. The RGBLED's (rgb_led) color is then set to a new Color created from the current red, green, and blue constituents. These are standard color values specified as red, between and blue components, each between 0 and 255.

Here is the code for the user interface itself, which is arranged in a grid:

```
app = App(title='RGB LED', width=500, height=400,
    layout='grid')

Text(app, text='Red', grid=[0,0]).text_size = 30

Slider(app, command=red_changed, end=255, width=350,
    height=50, grid=[1,0]).text_size = 30

Text(app, text='Green', grid=[0,1]).text_size = 30

Slider(app, command=green_changed, end=255, width=350,
    height=50, grid=[1,1]).text_size = 30

Text(app, text='Blue', grid=[0,2]).text_size = 30

Slider(app, command=blue_changed, end=255, width=350,
    height=50, grid=[1,2]).text_size = 30


app.display()
```

Note how the method text_size is used to make the text bigger. The command parameter of the Sliders links to the appropriate color change function.

Summary

This is a simple project to get you started with some GPIO programming. In the next chapter, you will use a display module that uses a I2C serial interface to connect to the Raspberry Pi and make a digital clock.

11

Prototyping Project (Clock)

In this chapter, we will build what can only be seen as a grossly over-engineered LED digital clock. We will be using a Raspberry Pi, a breadboard, and a four-digit LED display (see Figure 11-1).

In the first phase of the design, the project will just display the time. However, a second phase extends the project by adding a push button that, when pressed,

Figure 11-1 *LED clock using the Raspberry Pi.*

switches the display mode between displaying hours/minutes, seconds, and the date.

What You Need

To build this project, you will need the following parts. Suggested part suppliers are listed, but you can also find these parts elsewhere on the Internet.

Part	Suppliers
Adafruit four-digit seven-segment I2C display	Adafruit (Product 880)
Solderless breadboard	Adafruit (Product 64), SparkFun (SKU PRT-00112), Maplin (AG09K)
Jumper wires (male to male) or a solid core wire	Adafruit (Product 758), SparkFun (SKU PRT-08431), Maplin (FS66W)
Jumper wires (female to male)	Adafruit (Product 1954), SparkFun (PRT-09385)
PCB mount push switch*	Adafruit (Product 367), SparkFun (SKU COM-00097), Maplin (KR92A)
** Optional. Only required for Phase Two.*	

The breadboard, jumper wires, and switch are all included in the Project Box 1 Kit for Raspberry Pi by MonkMakes.

Hardware Assembly

The LED display module is supplied as a kit that must be soldered together before it can be used. It is easy to solder, and detailed step-by-step instructions for building it can be found on the Adafruit website. The module has pins that just push into the holes on the breadboard.

The display has just four pins (VCC, GND, SDA, and SCL) when it is plugged into the breadboard; align it so that the VCC pin is on row 1 of the breadboard.

Underneath the holes of the solderless breadboard are strips of connectors, linking the five holes of a particular row together. Note that because the board is on its side, the rows actually run vertically in Figure 11-2.

Figure 11-2 shows the solderless breadboard with the four pins of the display at one end of the breadboard.

Figure 11-2 *Breadboard layout.*

The connections that need to be made are listed here:

Suggested Lead Color	From	To
Black	GPIO GND	Display GND (second pin from left)
Red	GPIO 5V0	Display VCC (leftmost pin)
Orange	GPIO 2 SDA	Display SDA (third pin from left)
Yellow	GPIO 3 SCL	Display SCL (rightmost pin)

The color scheme shown in this table is only a suggestion; however, it is common to use red for a positive supply and black or blue for the ground connection.

CAUTION *In this project, we are connecting a 5V display module to the Raspberry Pi, which generally uses 3.3V. We can only safely do this because the display module used here only acts as a "peripheral" device and hence only listens on the SDA and SCL lines. Other I2C devices may act as a* *controller device, and if they are 5V, there is a good chance this could damage your Pi. Therefore, before you connect any I2C device to your Raspberry Pi, make sure you understand what you are doing.*

Turn on the Raspberry Pi. If the usual LEDs do not light, turn it off immediately and check all the wiring.

Software

Everything is connected, and the Raspberry Pi has booted up. However, the display is still blank because we have not yet written any software to use it. We are going to start with a simple clock that just displays the Raspberry Pi's system time. The Raspberry Pi does not have a real-time clock to tell it the time. However, it will automatically pick up the time from a network time server if it is connected to the Internet.

The Raspberry Pi displays the time in the top-right corner of the screen.

You may find that the minutes are correct but that the hour is wrong. This probably means that your Raspberry Pi does now know which time zone it is in. This can be fixed by setting the time zone from the Raspberry Pi Configuration tool that you will find in the Configuration section of the Raspberry Pi Menu (Figure 11-3).

By default, the I2C interface that the display uses is disabled, so before we can use the display we need to enable it by going to the Raspberry Pi Configuration tool in the Preferences menu and click on the Enabled button next to I2C (see Figure 11-4) and then click OK.

So now that the Raspberry Pi knows the correct time and the I2C bus is available, we can write a Python program that sends the time to the display. Adafruit has created some Python code to go with their I2C displays, in fact they have a very useful collection of all their Raspberry Pi code that is contained in a library called *blinka* that you need to download and install using the commands:

```
$ pip3 install adafruit-blinka
$ pip3 install adafruit-circuitpython-ht16k33
```

When you run the program `11_01_clock.py` the LEDs should light up and display the correct time.

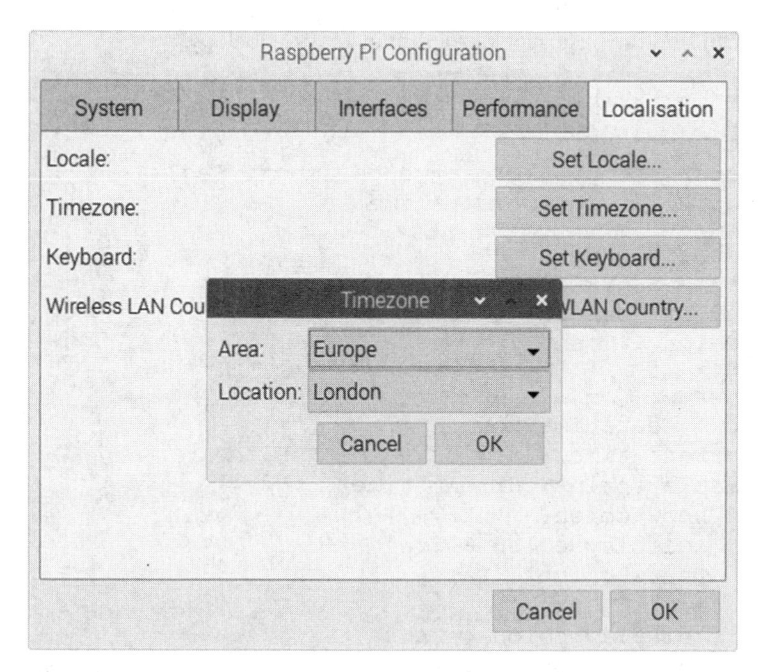

Figure 11-3 *Setting your Raspberry Pi's time zone.*

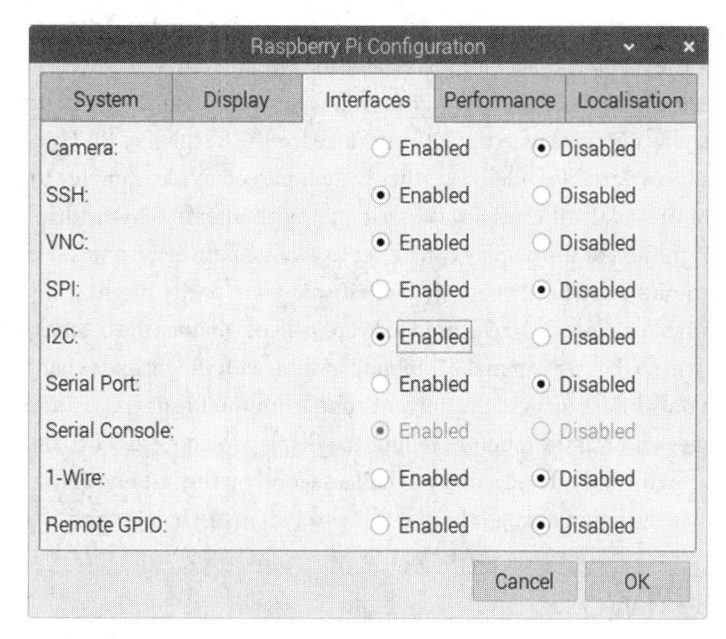

Figure 11-4 *Enabling the I2C interface.*

You can find the basic clock program in 11_01_clock.py.

```
# 11_01_clock.py

import board, time
from adafruit_ht16k33.segments import Seg7x4
from datetime import datetime

i2c = board.I2C()
display = Seg7x4(i2c)
display.brightness = 0.3
show_colon = True

while True:
    now = datetime.now()
    current_time = now.strftime("%H:%M")
    display.print(current_time)
    if show_colon:
        display.colon = True
        show_colon = False
    else:
        display.colon = False
        show_colon = True
    time.sleep(0.5)
```

The program starts by importing the things it needs from the Adafruit libraries to control the display. It also imports datetime which allows us to get hold of the date and time from the Raspberry Pi so that we can use it in the clock program.

The display modules uses the I2C interface for the Raspberry Pi. This interface is assigned to a variable called `i2c` that is then passed as a parameter to the constructor of the Adafruit class Seg7x4 that acts as an interface to the display itself.

The brightness of the display can be set to a value between 0 and 1. Here, it has been set to one-third brightness, as these displays are pretty bright.

The variable `show_colon` is used to keep track of whether the colon is currently being shown, so that we can make it reblink in time with the seconds changing.

The main while loop gets the current time from the Raspberry Pi and formats it into hours and minutes, before telling the display to show it. The colon is toggled between on and off and a delay of half a second on the last line of the program ensures that the colon changes between on and off every half second.

Phase Two

Having got the basic display working, let's expand both the hardware and software by adding a button that changes the mode of the display, cycling between the time

Figure 11-5 *Adding a button to the design.*

in hours and minutes, the seconds, and the date. Figure 11-5 shows the breadboard with the switch added as well as two new patch wires. Note that we are just adding to the layout of the first phase by adding the button; nothing else is changed.

NOTE *Shut down and power off your Pi before you start making changes on the breadboard.*

The button has four leads and must be placed in the right position; otherwise, the switch will appear to be closed all the time. The leads should emerge from the sides facing the top and bottom of Figure 11-5. Don't worry if you have the switch positioned in the wrong way—it will not damage anything, but the display will continuously change mode without the button being pressed.

Two new wires are needed to connect the switch. One goes from one lead of the switch (refer to Figure 11-5) to the GND connection of the display. The other lead goes to the connection labeled 18 on the GPIO connector. The effect is that whenever the button on the switch is pressed, the Raspberry Pi's GPIO 18 pin will be connected to ground.

You can find the updated software in the file 11_02_fancy_clock.py and listed here:

```
# 11_02_fancy_clock.py

import board, time, gpiozero
from adafruit_ht16k33.segments import Seg7x4
from datetime import datetime

switch = gpiozero.Button(23, pull_up=True)
i2c = board.I2C()
display = Seg7x4(i2c)
display.brightness = 0.3
show_colon = True
time_mode, seconds_mode, date_mode = range(3)
disp_mode = time_mode

def display_time():
    global show_colon
    now = datetime.now()
    current_time = now.strftime("%H:%M")
    display.print(current_time)
    if show_colon:
        display.colon = True
        show_colon = False
    else:
        display.colon = False
        show_colon = True
    time.sleep(0.5)

def display_seconds():
    now = datetime.now()
    current_seconds = now.strftime("  %S")
    display.print(current_seconds)
    time.sleep(0.5)

def display_date():
    now = datetime.now()
    current_date = now.strftime("%d%m")
    display.print(current_date)
    time.sleep(0.5)

while True:
    if switch.is_pressed:
        disp_mode = disp_mode + 1
        if disp_mode > date_mode:
            disp_mode = time_mode
```

```
    if disp_mode == time_mode:
        display_time()
    elif disp_mode == seconds_mode:
        display_seconds()
    elif disp_mode == date_mode:
        display_date()
```

The first thing to notice is that because we need access to GPIO pin 18 to see whether the button is pressed, we need to use the `gpiozero` library.

Most of what was in the loop has been separated into a function called `display_time`. Also, two new functions have been added: `display_seconds` and `display_date`. These are fairly self-explanatory.

One point of interest is that `display_date` displays the date in U.S. format. If you want to change this to the international format, where the day of the month comes before the month, change the format on line 36 to %d%m.

To keep track of which mode we are in, we have added some new variables in the following lines:

```
time_mode, seconds_mode, date_mode = range(3)
disp_mode = time_mode
```

The first of these lines gives each of the three variables a different number. The second line sets the `disp_mode` variable to the value of `time_mode`, which we use later in the main loop.

The main loop has been changed to determine whether the button is pressed. If it is, then `1` is added to `disp_mode` to cycle the display mode. If the display mode has reached the end, it is set back to `time_mode`.

Finally, the `if` blocks that follow select the appropriate display function, depending on the mode, and then call it.

Summary

This project's hardware can quite easily be adapted to other uses. You could, for example, present all sorts of things on the display by modifying the program. Here are some ideas:

- Your current Internet bandwidth (speed)
- The number of e-mails in your inbox

- A countdown of the days remaining in the year
- The number of visitors to a website

In the next chapter, we build another hardware project—this time a roving robot—using the Raspberry Pi as its brain.

12

Raspberry Pi Robot

In this chapter, you will learn how to use the Raspberry Pi with a motor chassis to make two versions of a roving vehicle. The first version allows the robot to be controlled using a web interface. The second version (Figure 12-1) is autonomous and will move around in a random manner, detecting obstacles in front of it using an ultrasonic rangefinder.

Figure 12-1 *A Raspberry Pi robot.*

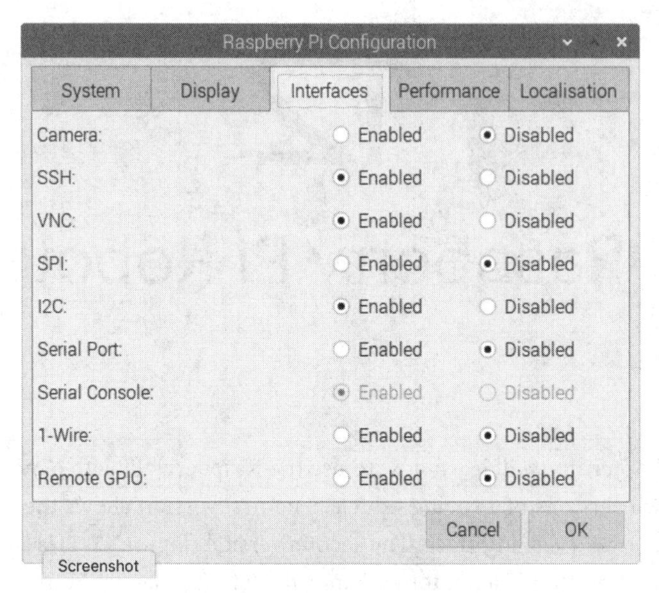

Figure 12-2 *Enabling the interfaces needed for this chapter's projects.*

In this chapter, I recommend the use of a Raspberry Pi Zero W (see Figure 12-2) for both projects. This half-sized Raspberry Pi is powerful enough to control the robot rover, but has the advantage that it uses a lot less battery power than say a Raspberry Pi 4 and also costs considerably less.

Set Up the Raspberry Pi Zero W

Although the Raspberry Pi Zero is a full-fledged Raspberry Pi, we are going to set it up to be *headless*. That is, once you have set everything up, you will be able to use it remotely from your computer without having to attach a keyboard, mouse, and monitor to it.

There are various ways of setting up a headless Raspberry Pi, but probably the most straightforward is to configure it with a keyboard, monitor, and mouse attached as normal and then allow it to go headless once it's configured.

So, start by using NOOBS to install Raspberry Pi OS, just like you did in Chapter 2. Make sure that you connect to Wi-Fi, as without this connection you will not be able to connect remotely to your Raspberry Pi when it goes headless.

There are a number of configuration changes that you need to make both to make the Raspberry Pi accessible remotely and so that it can use the Motor

Controller pHAT. Open up the Raspberry Pi Configuration tool and switch to the Interfaces tab. You then need to make sure that the options I2C, SSH, and VNC are all enabled (Figure 12-2) before closing the window.

I2C is the interface that the Raspberry Pi uses to communicate with the motor controller. This uses two GPIO pins to send commands to the motor controller.

SSH (Secure SHell) allows you to connect a command line tool (such as Putty on Windows or the Terminal on a Linux or Unix-based computer) to the Raspberry Pi, so that you can enter commands remotely from another computer as if you are using a Terminal on the Raspberry Pi itself.

VNC (Virtual Network Connection) is a graphical equivalent of SSH that allows you to see what you would see on your Raspberry Pi's monitor. Whereas SSH is text only, VNC allows you to interact with the Raspberry Pi using the keyboard and mouse of your computer. In this chapter, we will use VNC rather than SSH as it allows you to use the Raspberry Pi's graphical desktop, which can be useful when doing things like using the Raspberry Pi Configuration tool.

If you are an advanced user and prefer the command line, then you may prefer to use SSH rather than VNC.

While you still have your Raspberry Pi hooked-up to keyboard, mouse, and monitor, it's worth making sure that you can connect to your Raspberry Pi from another computer.

To do this, you need to find the IP address of your computer. Do this by entering the following command in a Terminal window:

```
$ hostname -I
192.168.1.56
```

So, in this case the IP address of my Raspberry Pi is 192.168.1.56. Note that this is an internal IP address. That means it is only accessible from another computer connected to your home Internet LAN (Local Area Network). Don't worry, your Raspberry Pi rover cannot be taken over by some malign Internet user.

Make a note of the IP address, as you are going to need it on a few occasions.

To be able to access your Raspberry Pi from a second computer, you need to download a VNC Viewer app onto your computer. The most popular free VNC viewer is available for most operating systems and can be downloaded from here: https://www.realvnc.com/en/connect/download/viewer/. Download and install it, making sure that you download the VNC Viewer app and NOT the VNC Server that is also available on that site.

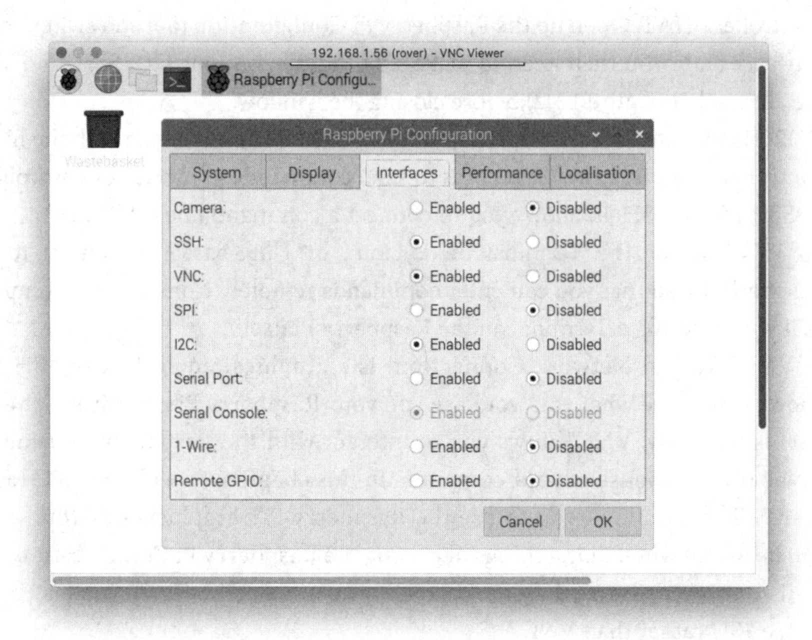

Figure 12-3 *Connecting to a Raspberry Pi over VNC.*

Open the app and in the address bar at the top of the window, enter the IP address of your Raspberry Pi (Figure 12-3) and then press Enter. You should then be prompted for your username (*pi*) and password (*raspberry*—unless you have changed it). After a few moments, a second window should appear on your computer that mirrors what you see on your Raspberry Pi's desktop and you have remote control of your Raspberry Pi.

Local IP addresses are not permanently allocated to a device when it connects to the network. This means that if we rebooted the Raspberry Pi or the router, we might find that the IP address changes. This would be inconvenient, as we would have to work out what the new IP address of our roving robot was, and made all the more tricky if our Raspberry Pi is headless. One way around this is to tell our network router (home hub) that the Raspberry Pi used on the rover is special and its local IP address should be allocated permanently. This feature is usually referred to as IP address reservation and you will find it by logging on to the admin console of your home router and looking for this feature. Routers have different admin interfaces and you may have to do a bit of Internet searching to find out how to do it on your particular router. On my router the interface looks like Figure 12-4.

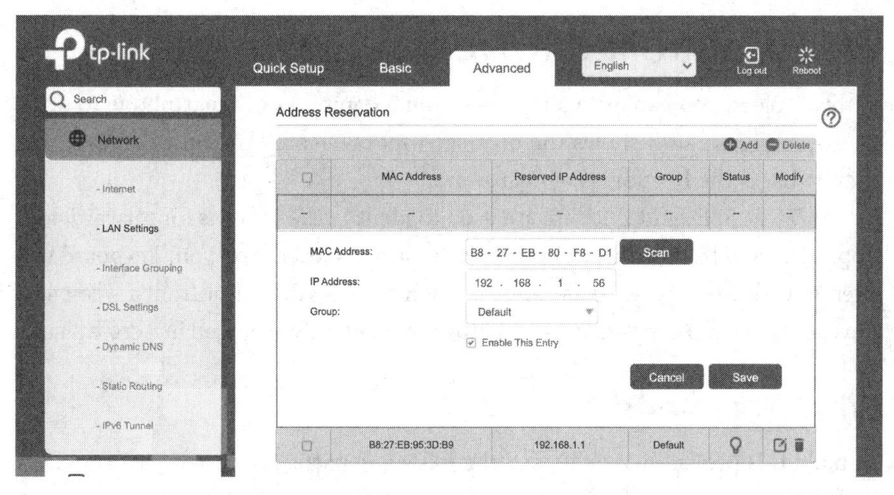

Figure 12-4 *Reserving an IP address for your Raspberry Pi on your router.*

If you are lucky, the interface might let you select your Raspberry Pi from a list of devices connected to your network. If not, then you might need to find the MAC address of your Raspberry Pi's Wi-Fi interface. You can find this by running the following command:

```
$ ifconfig wlan0
wlan0: flags=4163<UP,BROADCAST,RUNNING,MULTICAST>  mtu 1500
      inet 192.168.1.56  netmask 255.255.255.0 broadcast 192.168.1.255
      inet6 fd00::1:d4fc:6035:6395:1578 prefixlen 64 scopeid 0x0<global>
      inet6 fe80::e696:a71b:ee53:6570  prefixlen 64 scopeid 0x20<link>
      ether b8:27:eb:80:f8:d1  txqueuelen 1000  (Ethernet)
      RX packets 4314  bytes 427704 (417.6 KiB)
      RX errors 0  dropped 0  overruns 0  frame 0
      TX packets 3850  bytes 2693601 (2.5 MiB)
```

The MAC address will be next to the name `'ether'` and is shown in bold above for my Raspberry Pi (yours will be a different code).

Once you have the MAC address, it can be entered into the admin console of your router along with the IP address to be reserved as shown in Figure 12-4. From now on, whenever the Raspberry Pi connects to your network using the Wi-Fi interface, it should always be allocated this IP address.

Shut down your Raspberry Pi, disconnect the keyboard, mouse, and monitor and then power it up again and make sure that you can still connect from another computer using VNC.

Web-Controlled Rover

In this project, you can use your browser on a computer or smartphone to drive the rover. Figure 12-5 shows the browser window. Note how the address in the address bar is the IP address of my rover.

The W, A, S, D, and Z letters are used to identify the buttons for forward, left, stop, right, and backwards, respectively, because those keys on your keyboard will interact with the browser page, so that when steering the robot from a web page, you can use those keyboard keys which are conveniently arranged in a cross shape.

What You Need

To build this project, you will need the following parts. Suggested part suppliers are listed, but you can also find these parts elsewhere on the Internet.

Part	Suppliers
Raspberry Pi Zero W	–
Robot chassis (6V gear motors) with 4 × AA battery box	eBay, Amazon, etc.
5V USB battery pack	eBay, Amazon, etc.
Motor driver pHAT for Raspberry Pi	PiHut: 102606
Self-adhesive Velcro™ pads	Stationary store

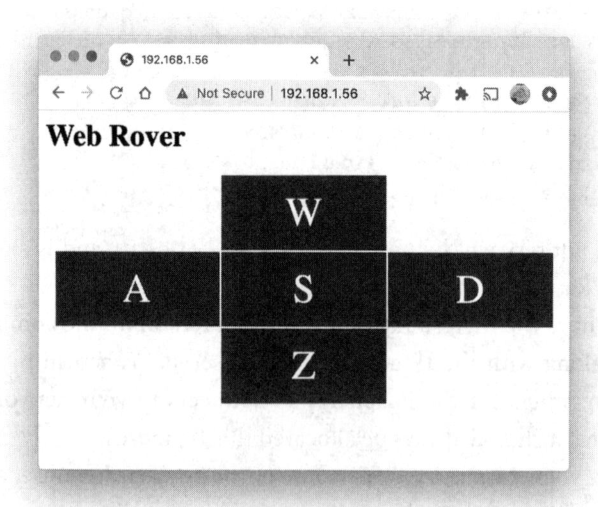

Figure 12-5 *A web interface for controlling your robot.*

Robot chassis are quite common on eBay. Look for something with 6V motors. The kits often come with a battery holder that accepts 4 × AA batteries.

The project uses this motor driver add-on for the Raspberry Pi: https://www.waveshare.com/wiki/Motor_Driver_HAT. In theory, this board is equipped to power your Raspberry Pi as well as your motors, but in practice, you will probably find that the motors will produce a sudden drain on the batteries when they start, which will cause your Raspberry Pi to reboot. So, the project will be more reliable if the Raspberry Pi is powered separately from the motors by using a 5V USB battery pack.

The self-adhesive Velcro is just a good way of attaching the Raspberry Pi and battery to the robot chassis in a way that it can still easily be removed if you need to.

Hardware

The motor chassis kits are all slightly different. Generally, by looking at a picture of the finished article, it's fairly obvious how to fit them together, although there may be a certain amount of trial and error. You may also find that the gearmotors that drive the wheels are supplied with wires that need to be soldered onto the motors.

Build the chassis, but leave the wheels off the gearmotors to stop the rover driving itself off your table while you are testing it out.

Power off your Raspberry Pi and fit the motor controller pHAT onto your Raspberry Pi being careful to line up the pins correctly.

Set the switch on the pHAT to off and connect the leads to the motors and the battery box as shown in Figure 12-6.

Note that during testing you may find that you have to swap over the motor leads for them to turn in the direction you would expect.

Software

Power up your Raspberry Pi. You can still use its regular power supply rather than the USB battery pack while you are getting everything working. It's also probably a bit early to fix the Raspberry Pi and battery box onto the chassis, but put some adhesive tape over any metal bolts on the top side of the chassis to prevent accidental short-circuits that could damage your Raspberry Pi.

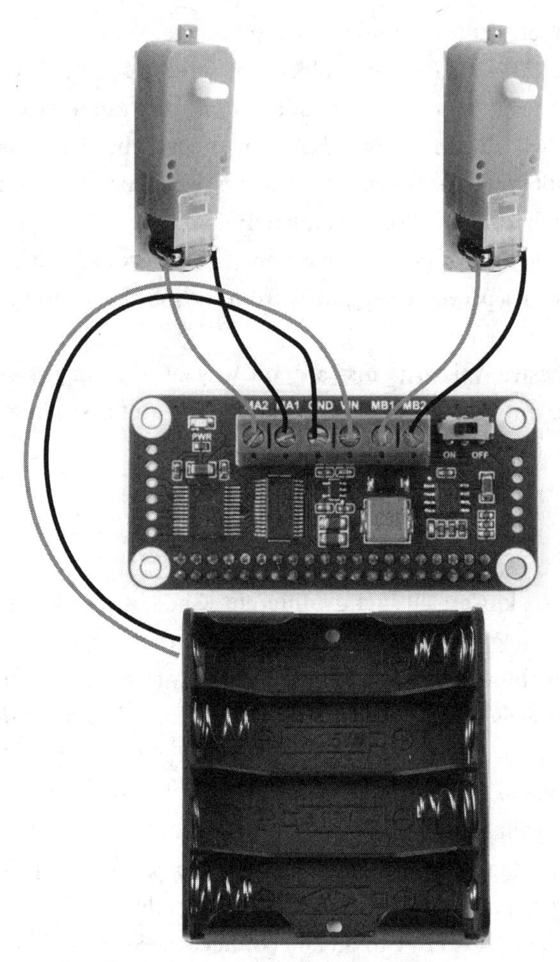

Figure 12-6 *Connecting the battery box and motors to the motor controller.*

Give your Raspberry Pi some time to boot up and then connect to it using VNC. From the VNC window on your computer, open a Terminal and run the following command to install the web server software (bottle) that this project uses:

```
$ sudo apt-get install python3-bottle
```

Navigate to the software for this project which you will find in mu_code/prog_pi_ed3/ch12.

Run the program using the following command:

```
$ sudo python3 12_01_rover_web.py
Bottle v0.12.15 server starting up (using WSGIRefServer())...
Listening on http://0.0.0.0:80/
Hit Ctrl-C to quit.
```

Now you can open a browser on your computer or smartphone (as long as it's connected to Wi-Fi) using the IP address of your Raspberry Pi, as shown in Figure 12-5.

Note that the program must be run using *sudo* because the bottle web server requires super-user privileges to be able to run a web server.

Let's have a look at the code:

```
# 12_01_rover_web.py

from bottle import route, run, template, request
from motor_driver_i2c import MotorDriver

motors = MotorDriver()

# Handler for the home page
@route('/')
def index():
    cmd = request.GET.get('command', '')
    if cmd == 'f':
        motors.forward()
    elif cmd == 'l':
        motors.left(0, 0.5) # turn at half speed
    elif cmd == 's':
        motors.stop()
    elif cmd == 'r':
        motors.right(0, 0.5)
    elif cmd == 'b':
        motors.reverse(0, 0.3) # reverse slowly
    return template('home.tpl')

run(host="0.0.0.0", port=80)
```

The program starts by importing various things that it needs from `bottle` to be able to act as a web server that will just serve a single page with the controls for moving the rover. The class `MotorDriver` is also imported from the module `motor_driver_i2c` which you will also find in the code download for Chapter 12 in the file motor_driver_i2c.py.

An instance of `MotorDriver` is created and assigned to the variable `motors`. The bottle framework allows you to specify a number of routes or web pages as I prefer to think of them. In this case, just one route (/) is specified using the @route directive. The handler function `index` will be called whenever a browser connects to the IP address of the Raspberry Pi.

The `index` handler function reads the `command` request parameter and sends different motor control commands depending on the request parameter. We will see in a moment where these request parameters come from, but continuing with the program, the last line of the `index` handler returns a template consisting of HTML contained in the file home.tpl. The `run` command starts the web server listening.

Here is a simplified version of the template file home.tpl with the code for intercepting key presses removed to keep things simple.

```
<html>
<head>
<script src="http://ajax.googleapis.com/ajax/libs/jquery/1.3.2/
jquery.min.js" type="text/javascript" charset="utf-8"></script>

<style>
.controls {
      width: 150px;
      font-size: 32pt;
      text-align: center;
      padding: 15px;
      background-color: green;
      color: white;
}
</style>

<script>
function sendCommand(command)
{
      $.get('/', {command: command});
}
</script>

</head>
<body>
```

```
<h1>Web Rover</h1>

<table align="center">
<tr><td></td><td class="controls"
onClick="sendCommand('f');">W</td><td></td></tr>
<tr><td  class="controls" onClick="sendCommand('l');">A</td>
    <td  class="controls" onClick="sendCommand('s');">S</td>
    <td  class="controls" onClick="sendCommand('r');">D</td>
</tr>
<tr><td></td><td  class="controls"
onClick="sendCommand('b');">Z</td><td></td></tr>
</table>

</body>
</html>
```

Some of this file is HTML to create the buttons on the web page and some of it is JavaScript code that is run when that button is pressed. For example when the S button is pressed, the JavaScript `sendCommand` function is called with "s" as a parameter. This results in a background web request being sent to the Raspberry Pi with a request parameter ("command") with a value of "s" (stop) that will then be handled by the Python `index` function, resulting in the motors being stopped.

If you are new to web interfaces, this can all be very confusing, because even though the file home.tpl is on the Raspberry Pi, the browser will download it, use it to display the web page, and run JavaScript all within the browser.

Watch which way the motor shafts turn when you press the control buttons. If it looks like one of the wheels is turning the wrong way (for the motion you commanded) then swap over the leads for that motor so that the red lead goes to where the black went and vice versa.

Once you are sure that everything is working as it should, it's time to fix down the Raspberry Pi and battery and put the wheels on the rover and set it loose!

Autonomous Rover

The autonomous version of this project is only autonomous in the sense that it will use an ultrasonic rangefinder to try and avoid getting stuck against obstacles.

What You Need

To build this project, you will need everything you needed for the first project, plus the following:

Half-sized breadboard	Adafruit: 64, SparkFun: PRT-12002
Ultrasonic rangefinder—HC-SR04**P***	Adafruit: 4007, eBay, Amazon, etc.
1kΩ 1/4W resistor (only if using HC-SR04)	SparkFun: PRT-14492, Adafruit: 4294
4 × female-to-male jumper wires	SparkFun: PRT-09140, Adafruit: 826
**Note that the HC-SR04P is a 3V version of the HC-SR04. This makes it easier to connect to a Raspberry Pi's 3V logic, but if you cannot find an HC-SR04P, an HC-SR04 and a 1kΩ resistor can also be used (see sidebar using a 5V HC-SR04).*	

The rangefinder works by sending out a pulse of ultrasound (40kHz) and then listens for an echo from any obstacle that the ultrasound bounces off. By timing how long it takes for the echo to come back and knowing the speed of sound, you can calculate the distance between the rangefinder and the obstacle.

Hardware

This project starts with the same wiring as the previous project, but now you also need to wire up the ultrasonic rangefinder as shown in Figure 12-7.

The solderless breadboard should have a peel-off self-adhesive base. Peel it back a little way to attach the breadboard where it can act as a base to hold the rangefinder where it can face forward to detect obstacles (see Figure 12-1) for reference.

There are two identical looking ultrasonic rangefinders on the market. There is the plain-old HC-SR04 that operates at 5V logic level and there is the HC-SR04P that operates at 3V. The latter is harder to find, so if you have an HC-SR04 and want to use it for this project, then you still can, but you will need a 1kΩ resistor and will also have to wire up the rangefinder as shown in Figure 12-8. The differences being that the HC-SR04 (NOT P) must have VCC connected to 5V rather than 3V and ECHO on the rangefinder must be connected to GPIO pin 18 via a 1kΩ resistor.

The HC-SR04 rangefinder uses two GPIO pins TRIG (trigger) and ECHO. The TRIG pin connects to a Raspberry Pi GPIO digital output and even though the output from the Raspberry Pi is 3V, it will still trigger the rangefinder to send a pulse of ultrasound without any extra circuitry.

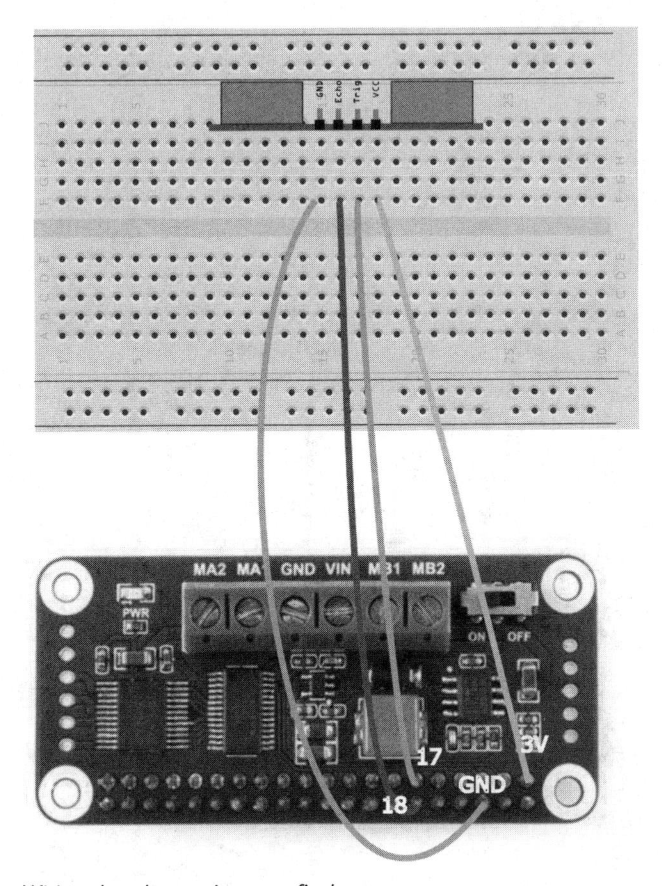

Figure 12-7 *Wiring the ultrasonic rangefinder.*

It's a different matter for the ECHO output of the rangefinder and because, on an HC-SR04, this is a 5V output, you would be connecting a 5V output to the 3V logic input of the Raspberry Pi's GPIO pin. If connected directly, too much current would flow into the input, causing heating that may immediately or over time damage your Raspberry Pi irreparably.

To guard against this, a 1kΩ resistor must be used in between the output of the HC-SR04 and the input of the Raspberry Pi. This will limit any current flowing to a safe couple of milliamps. The breadboard that supports the rangefinder is the obvious place to put the resistor.

Figure 12-8 *Using a 5V rangefinder with Raspberry Pi.*

Software

Once the hardware is complete, you can power up the Raspberry Pi. I would recommend taking the wheels off to prevent your robot's unexpected departure from your table. You will find the program for this project in mu_code/prog_pi_ed3/ch12/12_02_rover_avoiding.py. Open a terminal and run the program using the following command:

```
$ python3 12_02_rover_avoiding.py
56.639925234462396
56.416771908413125
56.33095690765418
55.11233241934848
19.600832519839173
```

```
18.879969356889433
18.879969356889433
18.948621356716103
/usr/lib/python3/dist-packages/gpiozero/input_devices.
py:997: DistanceSensorNoEcho: no echo received
  warnings.warn(DistanceSensorNoEcho('no echo received'))
19.223229356998388
19.429202519296894
19.377696357584
```

The numbers that appear are the readings from the rangefinder. Try holding your hand in front of the rangefinder and make sure that the rangefinder is working. Note that every so often you will see an error message saying that no echo was received. This is quite normal.

The basic strategy of the rover is that it moves forward until it detects an obstacle closer than 50cm, at which point it turns on the spot (by turning its wheels in opposite directions) for a random time before setting off again.

To stop the motors running, place an obstacle about 15cm (6 inches) from the rangefinder. The code is designed to stop the rover entirely if it gets too close to an obstacle.

Here's the code:

```
# 12_02_rover_avoiding.py

from gpiozero import DistanceSensor
from motor_driver_i2c import MotorDriver
import time, random

motors = MotorDriver()
rangefinder = DistanceSensor(echo=18, trigger=17)

def turn_randomly():
    turn_time = random.randint(1, 3)
    if random.randint(1, 2) == 1:
        motors.left(turn_time)
    else:
        motors.right(turn_time)
    motors.stop()

while True:
    distance = rangefinder.distance * 100 # convert to cm
    print(distance)
```

```
if distance < 20:
    motors.stop()
elif distance < 50:
    turn_randomly()
else:
    motors.forward()
time.sleep(0.2)
```

Very helpfully, the gpiozero library has a class called `DistanceSensor` that makes using an HC-SR04P super-easy. We just need to tell it which GPIO pins are connected to TRIG and ECHO and then we can ask it for a distance to any obstacle in meters using `rangefinder.distance`.

The `turn_randomly` function generates a random time between 1 and 3 seconds and then picks randomly between turning left or right for that amount of time.

Summary

These two robot projects can form the basis of lots of other interesting projects. You might decide to try and make a line-following robot or attach a USB webcam to it.

This is the final project in this book. In the next and final chapter you will learn about other resources and places to help you to program your Raspberry Pi.

13

What Next

The Raspberry Pi is a phenomenally flexible device that you can use in all sorts of situations—as a desktop computer replacement, a media center, or an embedded computer to be used as a control system.

This chapter provides some pointers for different ways of using your Raspberry Pi and details some resources available to you for programming the Raspberry Pi and making use of it in interesting ways around the home.

Linux Resources

The Raspberry Pi is, of course, one of many computers that runs Linux. You will find useful information in most books on Linux; in particular, look for books that relate to the distribution you are using, which for Raspberry Pi OS will be Debian.

Aside from the File Manager and applications that require further explanation, you'll want to know more about using the Terminal and configuring Linux. A useful book in this area is *The Linux Command Line: A Complete Introduction,* by William E. Shotts, Jr. Many good resources for learning more about Linux can be found on the Internet, so let your search engine be your friend.

Python Resources

Python is not specific to the Raspberry Pi, and you can find many books and Internet resources devoted to it. For a gentle introduction to Python, you might

- **http://elinux.org/RPi_Low-level_peripherals** A list of peripherals for interfacing with the GPIO connector.

If you want to go into more depth about the Raspberry Pi, especially connecting it to electronics, then you can refer to my book *Raspberry Pi Cookbook* which contains recipes for pretty much anything you might want to do with your Raspberry Pi.

If you are interested in buying hardware add-ons and components for your Raspberry Pi, Adafruit has a whole section devoted to the Raspberry Pi. SparkFun also sells Raspberry Pi add-on boards and modules. In the United Kingdom, CPC, Pimoroni, and Kitronik all sell interesting add-ons for the Raspberry Pi.

Programming Languages

In this book, we have looked exclusively at programming the Raspberry Pi in Python, and with some justification: Python is a popular language that provides a good compromise between ease of use and power. However, Python is by no means the only choice when it comes to programming the Raspberry Pi. The Raspberry Pi OS distribution includes several other languages.

Scratch

Scratch is a visual programming language developed by MIT. It has become popular in education circles as a way of encouraging youngsters to learn programming. Scratch includes its own development environment, like IDLE for Python, but programming is carried out by dragging and dropping programming structures rather than simply typing text.

Figure 13-1 shows a section of one of the sample programs provided with Scratch for the game *Pong*, where a ball is bounced on a paddle.

Simon Walters has developed a GPIO library so that you can use Scratch to control GPIO pins.

C

The C programming language is the language used to implement Linux, and the GNU C compiler is included as part of the Raspbian Wheezy distribution. To try

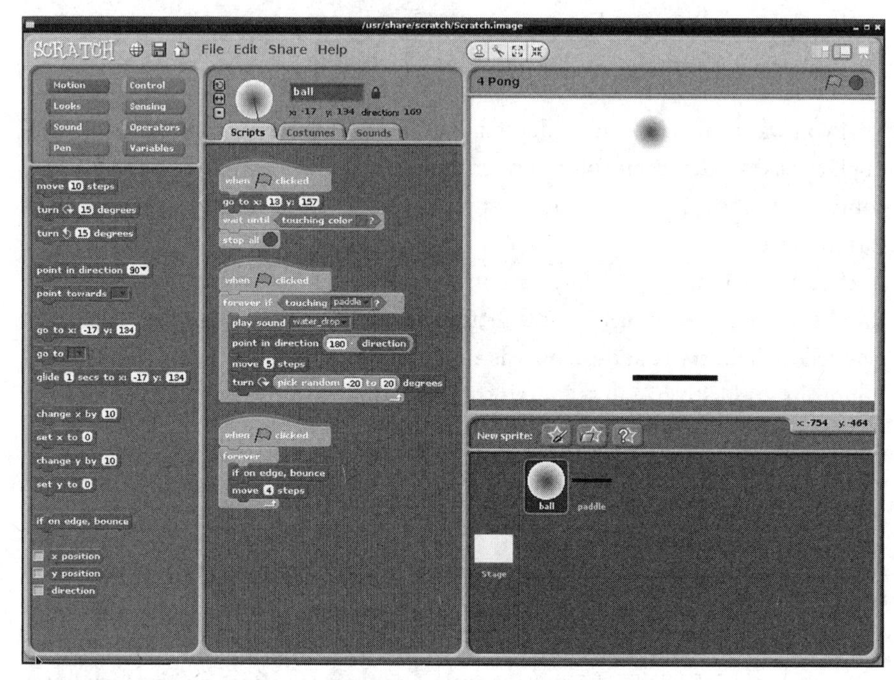

Figure 13-1 *Editing a program in Scratch.*

out a little "Hello World'" type of program in C, use Mu to create a file with the following contents:

```
#include<stdio.h>
main()
{
    printf("\n\nHello World\n\n");
}
```

Save the file, giving it the name hello.c. Then, from the same directory as that file, type the following command in the terminal:

```
gcc hello.c -o hello
```

This will run the C compiler (`gcc`), converting hello.c into an executable program called just `hello`. You can run it from the command line by typing the following:

```
./hello
```

The Mu editor window and command line are shown in Figure 13-2, where you can also see the output produced. Notice that the `\n` characters create blank lines around the message.

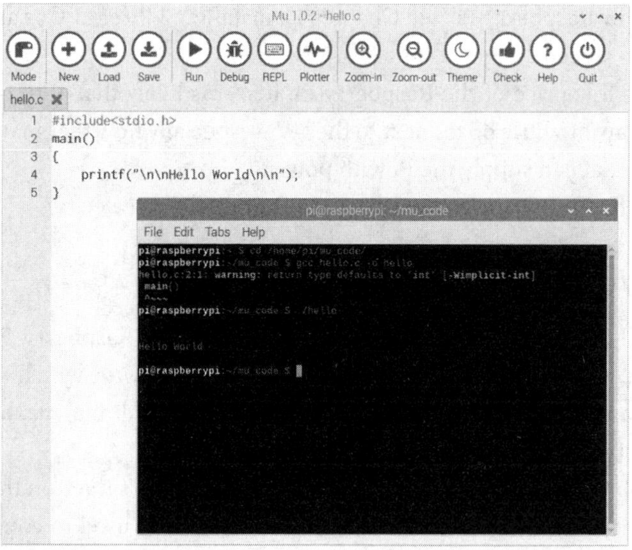

Figure 13-2 *Compiling a C program.*

Other Languages

You will also find a couple of IDE options for Java programming on the Start menu. So, if you want to learn Java (NOT to be confused with JavaScript), you can do so on your Raspberry Pi. The newest Raspberry Pi models are perfectly capable of running a web server stack, which includes a database and web server. The options for this change frequently with the latest fashion in web development.

Applications and Projects

Any new piece of technology such as the Raspberry Pi is bound to attract a community of innovative enthusiasts determined to find interesting uses for the Raspberry Pi. At the time of writing, a few interesting projects were in progress, as detailed next.

Media Center (Kodi)

LibreELEC is a distribution for the Raspberry Pi based on the Kodi software that turns it into a media center you can use to play movies and audio stored on USB media attached to the Pi, or you can stream audio and video from other devices such as iPads that are connected to your home network. When you use NOOBS to set up your Raspberry Pi (see Chapter 1), installing LibreELEC is an option on the NOOBS screen.

With the low price of the Raspberry Pi, it seems likely that a lot of them will find their way into little boxes next to the TV—especially now that many TVs have a USB port that can supply the Pi with power.

You can find out more about LibreELEC at https://libreelec.tv/.

Home Automation

Many small-scale projects are in progress that use the Raspberry Pi for home automation, or *domotics* as it is also known. The ease with which sensors and actuators can be attached to it, either directly or via an Arduino, make the Pi eminently suitable as a control center.

Most approaches have the Raspberry Pi hosting a web server on the local network so that a browser anywhere on the network can be used to control various functions in the home, such as turning lights on and off or controlling the thermostat.

Technologies to look for here are MQTT (mosquitto) and NodeRED.

Summary

The Raspberry Pi is a very flexible and low-cost device that will assuredly find many ways of being useful to us. Even as just a simple home computer for web browsing on the TV, it is perfectly adequate (and much cheaper than most other methods). You'll probably find yourself buying more Raspberry Pi units as you start to embed them in projects around your home.

Finally, don't forget to make use of this book's website (www.raspberrypibook.com), where you can find software downloads, ways of contacting the author, as well as errata for the book.

INDEX